Ferrets are available in a number of colors and markings.

A pair of albino ferrets.

FERRETS YEARBOOK

BY MARY FIELD

DEDICATION

To Mark, Dan, Marie,
Vincent and Wolfie
In loving memory of my
husband, Jay, and my
parents

QUESTION: What is less than five pounds in weight, adorable, cuddly, mischievous, larcenous, playful, cunning, amusing, affectionate, intelligent, and definitely smaller than a breadbox?
ANSWER: A ferret.

A litter of ferrets ready to go to their new homes. Ferrets, whose scientific name is *Mustela putorius furo,* are interesting, amusing, and playful pets.

INTRODUCTION

Ferrets (*Mustela putorius furo*) have a long history. They first appeared as domesticated pets of the pharaohs in Egypt for several years before being replaced by cats as the number one pet. They later appeared in Victorian England as both pets and working animals. Queen Victoria favored albino ferrets and was known to give these animals as rewards to chosen friends and diplomats.

HUNTING COMPANIONS

At that time ferrets were also companions to hunters in England. They were trained to flush out rabbits for the hunter by running through rabbit holes and forcing the rabbit to appear in the open—usually the rabbit's last earthly appearance. Today ferrets are still used as hunting companions in England, and while they are similar in appearance to the domesticated ferret, their temperament and odor are far different from their tamed cousins.

Ferrets were brought to the United States from England in ships, where they were useful in keeping the rat population under control. After the ocean voyage they were kept in cages which were hung in barns; the mere odor of an unneutered ferret is enough to discourage even the most persistent rodent. Naturally, the treatment these ferrets received did not encourage them to become the sweet, lovable pets they are today.

Spending quality time with your ferret will strengthen the bond between you and your pet.

FERRETS AS PETS

In the 1980s ferrets again achieved a place in the limelight. They were then termed the "hot pets of the '80s." This fame was both bad and good—the good being that many people became exposed to ferrets and came to love them and let them become part of their households. The bad part was that they were "fad" pets—adopted by owners who wanted the new "in" pet but did not care enough about them to learn about their care. Once the novelty wore off, these owners moved on to the next fad—I think it was the Vietnamese pot-bellied pig—and abandoned their ferrets.

A silver-foot, or silver-mitt, ferret. Ferrets continue to grow in popularity as more people discover their virtues as pets.

The 1990s brought about a different type of interest in ferrets. The fad aspect was over, and people who encountered and loved ferrets as pets were totally committed. They had seen the rewards of ferret ownership—ferrets are clean, odor free when neutered and properly cared for, occupy little space, don't bark or otherwise disturb the peace and don't need to be walked in inclement weather. In these times when most households are two-income, the owners can leave knowing that their ferrets are not pining away in loneliness; ferrets will curl up and sleep when they are not engaged in activity. It seems as if they are storing up energy so they can be super attentive when their owners return home. Many of the Ferret

Fanciers Club (FFC) members live in apartments where there is a strict no pet clause—they can enjoy pet ownership because ferrets are quiet and small—no one is aware they have these little pets unless the owners "confess."

Ferrets have become such an accepted part of life in the last few years that they have even been featured in such motion pictures as *Hollywood Cop*, *Legends of the Fall* and *Beastmaster*.

My late husband Jay and I acquired our first ferret, Furry, in 1984. We had seen a picture of a ferret in a magazine and decided a ferret would fit into our lifestyle. We lived in a "no pet" apartment and both worked. When we finally located a shop which could provide us with a ferret, we were dismayed to find there was no information on his care and no ferret-type supplies. We were forced to improvise. His first resting place was a round dish pan lined with towels. We placed it inside the bathtub in our guest bathroom. His first collar was manufactured for a small cat; we cut it down until it was approximately the right size for him. Since no ferret food was on the market, we took a guess (luckily it was right) and provided him with kitten chow. Poor little Furry—it's a miracle he not only survived but thrived.

Soon he was able to escape from the bathtub. We realized he could wriggle into impossibly small areas so, armed with cardboard, tape and plywood, we covered any hole or area where he could escape.

Three months after acquiring this unusual-but-wonderful pet, we were invited to a large party. Purely by coincidence we started talking with a man who revealed he had a very unusual pet—A FERRET! We were ecstatic—our hostess's food and drink were wasted on us—all we wanted to do was compare ferret tales (tails). On the way home we decided to try to contact other ferret owners to share experiences and acquired knowledge. We consulted the want ads on a daily basis—if someone placed a "ferret for sale" ad we would contact him or her and ask if they wanted to join our club. We acquired several members in the Pittsburgh area and a few months later were contacted by a reporter from our local paper. He was so entranced by ferrets that he wrote a long article, accompanied by a quarter-page picture of Furry. This attracted many other ferret owners in the area. A few months later, Associated Press ran an article about our ferrets (by that time we had adopted an abandoned ferret, Heather). After that we were bombarded with requests from ferret owners to join our club and receive our newsletter.

A ten-month-old sable female. Before you bring your ferret home, you should gather all the information that you can in order to start out right in ferret ownership.

THE GROWTH OF THE FFC

The Wall Street Journal did a front-page article on ferrets and the FFC in April 1986, and from thereon the Ferret Fanciers Club (FFC) rapidly increased its membership. It seems that many ferret owners felt the same need to communicate with others who shared their love of these little animals. Today the club has a membership in excess of 3,000 internationally and is still growing. This membership represents caring, loving owners who are totally committed to the care and welfare of these dear little pets. We have seen many advances in information, products and care available to ferret owners, and rejoice that this has happened.

Youngsters are especially attracted to ferrets. *Always be sure to closely supervise a young child when it is playing with a ferret, or any other household pet for that matter.*

Ferrets should be allowed out of their cages on a daily basis to ensure that they have the opportunity for exercise and play.

IS A FERRET FOR YOU?

Perhaps I have been a bit enthusiastic in my description of ferrets as pets—I'm sorry, I get a bit carried away. However, now I'm going to play devil's advocate and ask you to examine your motives for ferret ownership.

Actually, ferrets are not for everyone. FFC strongly advises against ownership of a ferret in a household which has a child under the age of six. Actually, no pet of any kind should ever be left with a child under that age without adult supervision. Ferrets are small and can easily be treated like a stuffed animal by a child. The ferret may try to protect itself by biting.

Are there people in your home who are not physically agile or who have vision problems? Again, ferrets may not be ideal pets since they can easily get under one's feet.

Do you frequently travel? While ferrets can be left alone overnight on a rare occasion, they do thrive on human companionship and should have at least two to three hours a day of contact with their owners. If they are kept in a cage (more about that later), they do need time

When handling a ferret, it should be firmly supported from underneath. Your pet shop dealer can teach you how to properly handle your ferret.

This cat and this ferret have been raised together since they were young and enjoy each other's company. How well a ferret gets along with a cat depends largely upon the individual personality of the ferret involved.

Ferrets are very active pets whose antics amuse and delight their owners.

out for exercise and socialization. If you are not able to give a ferret the time and freedom it requires, it's best to look into less demanding pets such as fish or some types of cage birds.

Ferrets are not inexpensive—either to purchase or maintain. After the initial cost of the pet and supplies, they will need preventive immunization, annual check-ups, high quality food and good cages for housing. Are these costs in your budget?

If you still feel a ferret will fit into your home, plan ahead for its arrival.

Ferrets love to explore nooks and crannies. You can provide your ferret with a variety of items that will safely let him do this.

PREPARING FOR YOUR FERRET

Since you are still reading at this point, I assume you have determined that a ferret will indeed fit into your household. The next step is to acquire the ideal one. The FFC receives many questions regarding the best place to buy a ferret. Some people feel that breeders are the best providers. While that may be true, the FFC is unable to recommend retail breeders. Breeding ferrets is a complicated business; since we are unable to investigate and approve small breeders, we do not recommend any of these facilities. Good pet stores get their stock from large, well-run breeding farms. These ferrets are raised to be good pets. If a ferret of poor quality is bred, the offspring (kits) will not be satisfactory. Also, to become a pet stock ferret, a "socialization" process must be completed before separation from the mother. If your local pet store does not have a supply of ferrets (and this can happen frequently since ferrets are in great demand), we suggest that you place an order and ask the store owner to call you when a new shipment arrives.

INITIAL PREPARATIONS

Before you buy your ferret, there are a few preparations to be made at home. Decide which parts of the house will be available to the ferret. Ferrets are happy and secure in a large, comfortable cage at night and when you are not home with them. However, they should be permitted at least two hours a day freedom to roam and play with family members. Any rooms which will be available to your ferret must be thoroughly "ferret proofed."

Ferrets can and do get into impossibly small spaces. Any potential trouble spot should be covered with cardboard held in place with duct tape. If the kitchen is part

Long and lithe aptly describe the conformation of a ferret. These qualities enable a ferret to slip into the most unlikely places, so always keep an eye on your pet when it is out of its cage.

Your ferret's bedding should be changed on a regular basis. Pet shops carry suitable bedding material for ferrets and other small pets.

of the ferret's range, consider blocking the area leading to the back of the refrigerator and stove with plywood. Be sure there is no way the ferret can get into any cabinets, especially those containing cleaning products or detergent.

If the room you want to use for your ferret does not have a door, try this type of barrier. Use two pieces of wood 1" X 1" X 2' long with a $^1/_2$" deep and slightly more than $^1/_4$"-wide groove centered on the 2' side and cut the length of the wood. Attach one piece on each side of the doorway. Use a 2' high piece of $^1/_4$" plywood cut wide enough to slide in the grooves. Another method is to nail a strip of corner molding to each side of the door frame and molding to hold the barrier wood. Then slip a sheet of thin wood (as wide as the door frame and as tall as needed) down the tracks. The solid wood sheet is unclimbable, sturdy, and is easily stepped over by humans. The barrier is simple to remove when not needed.

Toys made of soft rubber are *not* suitable for ferrets. A ferret may chew on them and ingest bits of rubber, which can cause an intestinal blockage.

A three-month-old silver-mitt ferret. Another fun activity for ferrets is tunneling in and out of various objects, such as the piece of pipe shown here.

Are all window screens secure? If you have a slot-type mailbox, be sure the ferret cannot climb up and out of it. There have been several ferrets who escaped in this manner. If your washing machine and dryer are in the room, NEVER leave the doors open—ferrets love to sleep in a warm area and the dryer may just fill the bill. To be safe, never use the washer or dryer without checking them first. Reclining chairs are a grave danger to ferrets. They can easily crush a ferret to death. If you have one, it's best to restrict the ferret from the room where it is or try this tip: put two aluminum pie pans, turned upside down, under the chair. This will prevent your pet from lying down under the chair. The pans will act as an alarm if your ferret walks on them.

Owning two ferrets can be twice the fun as owning one, and they can keep each other company when you are not at home.

A cardboard box is fine to be used as a playhouse, but it is not suitable as permanent housing for a ferret.

Seemingly innocuous substances and situations can be fatally disastrous for ferrets. These include open toilet bowls, bathtubs with even one inch or less of water, a variety of houseplants with leaves to chew upon, wet floor wax, shoe polish to lick, household cleaners, inhalation of drain cleaner or household bleach, electrical wires to nibble on, mothballs and rodent control items. Once the house is ready, it's time to shop.

A ferret that is kept singly will amuse itself with its toys. Provide an assortment of toys that are safe and that your pet will enjoy.

CHOOSING YOUR FERRET

When you find a pet shop with a selection of ferrets, how do you choose? Granted, it's difficult since they all look so cuddly and precious. However, now it's time to be objective. Since a ferret's lifespan is eight to ten years, you and your pet will be together for many years and you don't want to err at this point.

Be sure the pet shop is clean and odor free. While ferrets do sleep a good bit while not being played with, after they are roused they should be alert and inquisitive. Check out their coat. It should be shiny, soft and completely covering the body—no bald spots. Examine the animal for any sores or scaly spots.

Pet stores sell neutered ferrets for a very good reason. An unneutered male is not a good pet since he emits a strong odor and is aggressive. A female MUST be neutered because if she goes into heat and is not bred, she will develop aplastic anemia and face a sure and agonizing death.

For reference, baby ferrets are called *kits*, females are called *jills*, and males are referred to as *hobs*. Are you uncertain whether to purchase a hob or a jill? If a ferret is neutered, there is no difference in temperament. The hob will be slightly larger (3-5 pounds) while the jill is 1-$^{1}/_{4}$ to 3 pounds. A hob's head is round and broad with a short tapered nose while the

This ferret is captivated by its own appearance in the mirror. A sound, healthy ferret will be alert and interested in its environment.

A litter of four-week-old youngsters. The average life span of a ferret is between eight and ten years.

head of a jill is smaller with a slightly longer tapering toward the nose.

Ferrets come in several colors. The color is purely a personal preference and makes absolutely no difference in the temperament or health of the ferret. Since ferrets have achieved popularity as domestic pets, many color patterns have been recognized with standards for each.

COLOR VARIETIES

SIAMESE—light-colored animal with legs, mask and tail showing darker complimentary color.

SABLE POINT—undercoat white to cream. Guard hairs and points dark brown.

CHOCOLATE POINT—undercoat white to cream, points and guard hairs milk chocolate.

BLUE POINT—undercoat white to blue-white, guard hairs and points slate blue.

RED POINT—undercoat white to buff, points and guard hairs from red to rust to reddish brown.

LILAC—undercoat white to blue-white, points pinkish or lavender gray.

ALBINO—solid white. Sexually mature adults may have a yellowish tinge.

BLACK SELF—solid black.

RED SELF—Irish setter red.

SABLE—undercoat white to cream. Points, guard hairs and belly fur dark brown.

BLUE—undercoat white to blue-white, points and guard hairs slate blue.

RED—undercoat buff to orange. Points and guard hairs orange to red.

MAUVE—undercoat white to cream. Points and guard hairs pinkish tan.

SILVER—undercoat white to cream. Points silver to silvery tan, guardhairs tipped with gray.

SILVER MITT—any color except albino or white. May have white paws, body markings, shields, tipped tails, blazes, etc.

BI-COLOR—any color and white distributed in a 50/50 ratio.

New colors are being continually developed since breeders feel the need to

An albino ferret. In some adult specimens, the fur may not be pure white.

A playful group of ferrets. Ferret personality can vary, even among members of the same litter.

branch out. Again, we urge you to choose the color you feel an attraction to—a ferret which has been raised by a breeder who adheres to strict standards will be a fine pet.

Check the ears. They should be pink and free of debris. The nose should feel moist; however, if your ferret has been cuddled into a towel while sleeping, the nose will feel dry for several minutes. The paws and nose should not be scaly or cracked.

There may be a slightly unpleasant odor to the ferret, since it has probably been cuddling with other ferret kits. If the ferret has already been neutered (as is usual with ferrets obtained in a pet store) the odor will disappear with a bath.

FERRET SUPPLIES THAT YOU WILL NEED

While you are at the pet store, this is the time to pick up supplies. There are now several good ferret foods on the market. We urge you to buy a supply of the food the ferret has been eating because it's a bit stressful for a ferret to come to a new home, and familiar food will make matters easier. If you decide to switch foods, do it gradually. After feeding the accustomed food for five-six days, mix it with the new food you favor. The first mix should be $^3/_4$ familiar food—$^1/_4$ new food. The next day increase the ratio of new food, and within four days the ferret should totally accept the new food.

Also, buy a ferret collar. Be sure the collar fits securely to prevent accidents and strangulation. Bells are a good investment—they will help you find your pet. A cage is best to house your ferret while you are out or asleep. The best cage is a two-story, with outdoor, washable carpeting on the floor. If the floors are left uncarpeted, the feet will be harmed from walking on the wire.

At this time it's a good idea to buy a biting deterrent (to discourage bad behavior) and a tasty supplement that your

When choosing a ferret, pick the color that most appeals to you, but remember that the animal's overall health should be your most important consideration.

A healthy ferret will have bright clear eyes and nice glossy fur with no thin or bare spots.

A lovely silver ferret. Male ferrets are called hobs; female ferrets are called jills.

pet shop dealer can recommend. A carrying cage is handy if you plan on traveling with your ferret—it's also good for the trip home or visits to the veterinarian. One designed for a cat will be more than adequate. If the carrying cage is not to be purchased at this time, take a small cardboard box with crushed paper towels in it for the ferret to cuddle up in for the trip home.

A water bottle for the side of the cage is best. Ferrets tend to tip water dishes.

Buy a small, heavy food dish that is not easily tipped.

A litter box should be placed in the first floor of a two-story cage. This might be an aluminum or plastic pan that is easily washed and disinfected. Place no more than two inches of gravel-type (dust-free) kitty litter in the pan. Ferrets are easily litter trained, but it takes a bit of patience.

Ferrets love a place to cuddle in and hide. Always provide soft, clean towels for this purpose. Also, a tube (a good one is the vent pipe from a clothes dryer) provides hours of fun and a secure place to sleep.

A squeaker-type toy made of hard rubber provides more than entertainment. Ferrets are easily trained to respond to the noise of the toy and will be coaxed out of hiding places when summoned with this toy. It has proven to be a true lifesaver (literally) in many instances.

To train your ferret to respond, squeak the toy one or two times. The noise will attract your pet; when he/she comes to you, reward with praise and a treat. Repeat this a few times daily for several months, and it will become an automatic process.

Are two ferrets better than one? Let's buy one initially and then consider a buddy for your pet.

When transporting your ferret, keep its carrying box away from a direct blast from the heater or air conditioner.

Don't let your ferret chew on just anything! Pet shops stock toys and other items that you can give to your pet.

These ferrets were socialized at an early age and are tractable and even tempered. Not all ferrets are this well behaved.

HOMECOMING

Your ferret has had a big day—it's exhausting coming from a pet store to a new home. The car ride, the move from a familiar cage with other ferret pals to a new, solitary cage, the loving and cooing from a new family, as nice as it is, can be tiring for a little pet. This is definitely not the day to take your new companion to visit friends or invite callers to see the new addition to the family.

When you bring your ferret home, give it time to become acclimated to its new environment. It won't be long before it becomes a member of the family.

When you reach home, put your ferret in his/her new cage which has been provided with a soft towel for cuddling. Allow the new arrival time for a nap.

When he/she wakens, permit time to use the litter pan. Don't remove the ferret from the cage until the litter pan has been used. If your ferret starts to run around the cage or back into corners, pick him/her up and place gently in the pan. When the pan has been used, then it's cuddle time.

Ferrets should be held gently but firmly. When getting to know your ferret, speak to it softly and give it lots of loving pets and cuddles. Love with a ferret might be thought of as an investment—it seems the more affection and love they receive the more they return.

BITING

A ferret kit is not unlike a puppy—it wants to bite. Ferrets have thick skin and bite their littermates in play. The kit doesn't know you don't have skin that will withstand a hard bite, but he/she must be taught. If you are bitten too hard for comfort, say "No" in a loud, firm voice and put the ferret down and ignore it for five minutes. If biting persists, follow these actions with a spray of biting deterrent in the face. Be consistent and the biting stage will pass.

Yes, ferrets *do* bite. Never tease your ferret or encourage "play-biting." If your ferret does bite, firmly scold it and be consistent in your training. In the case of a persistent biter, it may be necessary to use a sprayable biting deterrent.

Do you have other pets? Ferrets get along well with dogs and will either love or ignore your cat. Just remember, don't leave the animals alone unsupervised until you are POSITIVE they will not harm each other.

Ferret housing should be roomy, sturdy, and easy to maintain.

These ferrets are about to be let out for playtime. When they are finished playing, they should have easy access to their litter pan.

Ferrets enjoying their cuddly menagerie. If you are giving your ferret free access to various areas in your home, you should be prepared for toilet accidents that may occur.

LITTER TRAINING

Ferrets are easily litter pan trained. Be sure never to remove your pet from the cage after a nap until the litter pan has been used. If your ferret is roaming around his/her permitted areas, watch for any signs of running about nervously or backing into a corner. This is the signal the pan should be used. Place litter pans in the corners of the area where the ferret can roam. Ferrets will use the pan willingly if it is convenient enough, but they won't go to great lengths to find a pan.

Ferrets are very fastidious (some more than others). If the litter is soiled or needs to be changed, they will shun the pan. Clean the feces from the pan regularly, rake through the litter to keep it fresh and change the litter frequently. However, in the early stages do not remove all the feces from the pan. A little bit left there reminds the ferret of the purpose of the pan. Also, completely fresh litter seems like a wonderful sand box for the ferret to play in.

When pouring the litter in the pan, you may see dust rise. This dust can cause complications if inhaled by ferrets. To avoid this, pour the litter in an area away from ferrets. Sprinkle a few drops of water over the fresh litter to settle the dust before putting the pan down for the pet's use.

If your pet absolutely insists on using the wrong corner, simply put the pan in the chosen corner. Sometimes it's easier to go along with the ferret's wish.

One of our members reports he solved the problem of using "unauthorized corners" by placing his ferret's favorite sleeping blankets in the chosen corners. Because ferrets try to avoid soiling their living quarters, in a short time the problem was solved.

If your ferret is using a carpeted corner, try spraying the area with rubbing alcohol. Ferrets detest the odor and will avoid these areas. (Test the alcohol on a hidden area of your carpet first to be sure it will not leave a stain).

Incidentally, carpets stained by ferret "errors" can be cleaned with this simple and effective carpet solution. Mix one part white vinegar and one part powdered detergent with two parts water. Apply with a light circular motion to spot with white cloth, blot up and rinse with clear water. Repeat if necessary. Try on an inconspicuous spot of carpet first.

Patience and persistence are very important when it comes to litter training your ferret. Also, remember to keep the litter pan clean; otherwise, your pet may choose to "do his business" elsewhere!

Sometimes, a ferret may try to use a houseplant as its litter pan. To counteract this problem, put the plant out of the ferret's reach.

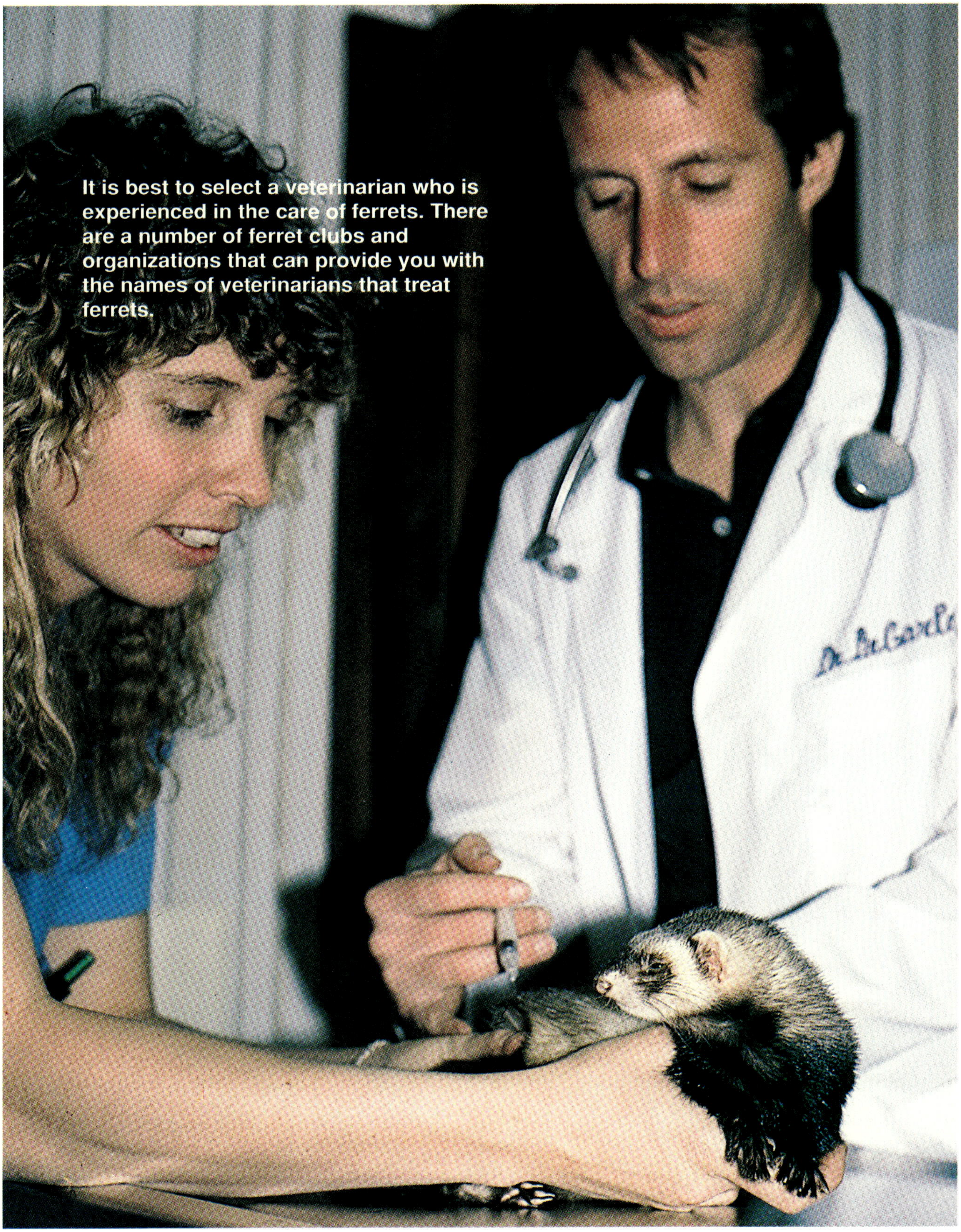

It is best to select a veterinarian who is experienced in the care of ferrets. There are a number of ferret clubs and organizations that can provide you with the names of veterinarians that treat ferrets.

MEDICAL CARE

Within the first week of ferret adoption, schedule the first visit with a veterinarian. Not all veterinarians are familiar with ferret care—it's best to ask before scheduling an appointment. If you are unable to find someone knowledgeable about ferret care, contact the FFC. We have a list of veterinarians across the United States who are trained in this area.

THE INITIAL CHECK-UP

The initial check-up should include an ear mite check, a canine distemper and rabies vaccine and any other immunizations your veterinarian feels are appropriate. Bring a stool sample to the initial visit so a check can be made for parasitic worms. Ask any questions about nutrition, care, etc. and arrange for neutering or spaying if needed.

This is a good time for the veterinarian to get to know your pet—that way if there is a problem the doctor will already be familiar with the ferret. Inquire if the veterinary clinic has night or emergency care—if not ask to be given the number of such a facility.

You will quickly come to know your ferret's daily habits, and if they begin to vary greatly, it is time to call your veterinarian. Some things to watch for are: lack of appetite, listlessness that lasts for more than one day, continual ear scratching and unusual odors.

Follow the recommendations of the veterinarian on vitamin supplements, frequency of immunization and other health care matters.

ADMINISTERING MEDICINE

If your veterinarian prescribes medication, you will find that administering medicine to a ferret is not a fun experience for either owner or pet, but one of our members shared the following tips for making it much easier:

1. Liquify a pill in a small oral syringe rather than hiding it in food or trying to force it down the ferret's throat.
2. Try to medicate the ferret when he's already awake, rather than forcing him/her to waken from a deep sleep.
3. It's best to give oral medicine when the ferret has some food in his/her stomach. If given on an empty stomach it may be thrown up.
4. If more than one kind of medicine is given, try to do

The more familiar you are with your pet's daily habits, the easier it will be to determine when he is not feeling well.

Today there is a variety of health and nutrition products formulated especially for small mammals such as ferrets.

all the medicating at the same time. If possible, mix all together.

5. Wear a full apron or protective smock to prevent clothing damage.

6. Hold the ferret in your lap on your left arm, slide syringe into side of ferret's mouth and push the plunger in quickly.

7. Try not to allow syringe to touch whiskers. This seems to alert the ferret medicine is on the way and they struggle.

8. Follow the medicine with a good-tasting treat and then play with your ferret. This will make your pet forget the unpleasant experience.

Above: **A scale can be the perfect place to curl and take a nap. Ferrets reach their adult weight between the ages of four and six months. Males average around three pounds; females average about one and one-half pounds.** ***Below*****: This ferret gets a clean bill of health from the vet. Regular veterinary care is part of responsible ferret ownership.**

NUTRITION

This section is based on information provided by Thomas R. Willard, Ph.D., an animal nutritionist. Dr. Willard has over 24 years' experience in the pet food industry.

Selecting the correct food has great impact on your ferret's health, overall well-being and your peace of mind. Poor diet can predispose a ferret to poor health and, in extreme cases, cause nutritional diseases. A well-balanced, quality ferret food will build and maintain an effective immune system, a strong muscle and skeletal framework, and properly functioning internal organs.

Ferrets are carnivores and, as such, require high quality animal source proteins and fats daily. This food must be fresh in order to attract the ferret, nutritious, and efficiently utilized by the body.

NUTRITIONAL REQUIREMENTS

The ferret requires seven nutrients, which are as follows:

Protein: Made of individual building blocks called amino acids. Ferrets require at least ten of these amino acids. Excellent sources of amino acids for ferrets are chicken and meat by-products, eggs, fish meal, and liver.

Fat: Required by the ferret daily as a primary source of energy, which helps support its active lifestyle. Animal fats are more digestible than vegetable oils and enhance the overall palatability of the food. Fats are made up of individual units called fatty acids. These fatty acids keep the skin supple and the haircoat thick and soft. Generally, poultry fat is the highest quality source of fats for ferrets and provides a taste which they prefer. Some premium-grade animal fats from pork or beef can also be used along with fish and vegetable oils for balance of the essential fatty acids.

Simple Carbohydrates: Starch is necessary in a quality ferret food to properly manufacture and shape it. Premium ferret foods are all cooked in a process known as extrusion. In this process, the starches are hydrated and heated to over 200° F., which completely cooks them. When cooled, the starch produces the crunchy texture and shape preferred by the ferret. This process also provides the structure to hold all the other nutrients together until it is eaten and digested by the ferret. Foods that do not have the starches properly cooked are poorly digested and usually are not palatable to the ferret. Loose stools are a common indication of this type of food processing.

Complex Carbohydrates: The fibrous portion of a food is extremely important in the daily diet of the ferret. Unlike other carnivores, the ferret does not have a caecum that aids in the digesting of this fiber. For this reason, feeding high fiber diets—those with 3% or more as in all dog and cat foods—can cause a digestive as well as a stool problem for ferrets. Feeding a food with 2% less fiber is very critical to good digestion and stool formation in the ferret.

Vitamins: Unlike proteins, fats, carbohydrates and certain minerals, vitamins are required in very small quantities. A premium ferret food will list 15 or more vitamins on the ingredient panel and should never require additional

Providing your ferret with a balanced, nutritious diet is essential to its overall health. Ferrets have a very fast rate of metabolism, and the food that they eat is utilized for body functions very quickly.

supplementation of vitamins or any other nutrients. Though excess vitamins can be tolerated by ferrets for a short time, longer term could cause metabolic problems that may lead to health problems.

Minerals: Mineral level and balance is most critical. Imbalances between the essential minerals can cause major problems and initiate many diseases and debilitating abnormalities in the ferret. Mineral supplementation of a food should not be attempted unless directed and supervised by a knowledgeable nutritionist or veterinarian. A premium ferret food will have all the essential minerals in the correct proportions, which will be listed in the ingredient statement of the food.

WATER

Water is the most important nutrient in the ferret's daily diet. It must be provided constantly or the ferret will rapidly dehydrate and die within a short period of time. Water must be from non-contaminated sources and be changed regularly.

A ferret that gets regular playtime and exercise should have a good appetite.

SELECTING FERRET FOOD

When selecting a ferret food, follow this guide. Look for a food that has been developed for and tested on ferrets. The label should state it is complete and balanced for growth, showing, gestation, lactation or maintenance. If the package does not specifically say it has been tested, don't use it! Call the manufacturer if you have any questions.

Pet shops stock dry food that is specially formulated to meet a ferret's dietary requirements. Photo courtesy of Marshall Pet Products.

Some veterinarians recommend a premium cat food for ferrets. This is often too high in fiber and low in many essential vitamins and minerals needed for the ferret.

Many cat foods have vegetable proteins, which are poor-quality ingredients for ferrets. A premium ferret food that is good for all life stages should be at least 36% protein and 22% fat with 2% or less fiber. Older ferrets also do well on this type diet but can become overweight if their activity level greatly decreases. Lower protein foods ranging

Your ferret's food should be of top quality. It should contain all of the important nutrients needed for growth and maintenance. Never sacrifice quality for quantity.

from 31-33% and 16-18% fat may be better suited for them. If this food cannot be found, then feed smaller portions of higher protein food. For a recuperating ferret, generally the higher protein foods should be fed. The higher level of nutrients helps the ferret recover more quickly because it promotes faster healing. This should always be discussed with and supervised by your veterinarian.

When selecting a food for your ferret, you should choose the best food you can afford. In general, premium ferret foods, high in protein and fat, will require less food to maintain the animal's weight, promote optimum growth and provide for the ferret's well-being. Since less food will be eaten, stools will be smaller and there will be less odor. When a palatable food is fed, there is less waste at the feeding dish, and overall feeding cost is lower than when feeding cheaper, less quality products.

Higher quality foods need to be packaged in a double-lined bag which keeps the food

A mother ferret and her youngster. In general, ferrets are not finicky eaters, and providing a diet that is both nutritious and appealing is a fairly straightforward matter.

These young ferrets have been weaned and are starting to eat solid food. The food has been moistened to a mash-like consistency to make chewing easier. In a few days, they will be able to eat the food completely dry.

A ferret that has been properly socialized will enjoy contact with humans.

fresher. Plastic bags often have small pin holes that allow air to get into the food and cause the product to become rancid.

It's amazing the amount of food such a small animal can consume. This is due to their high metabolism rate. We recommend food specially formulated for ferrets, which can be found in pet stores. If your kit is having trouble eating the hard food, you might soak it in water for a week or so. After that, hard food is beneficial to teeth and gums.

Besides basic food, ferrets enjoy all types of snacks. There are some definite NO-NO's in a ferret's diet. Milk and dairy products can cause diarrhea. Chocolate and sweets should be avoided as should alcohol in any form. Cooked meat, chicken, liver, peanut butter, cheese and eggs are nutritious goodies. Ferrets' taste in snacks is amazing—what one loves another will detest. Popular snacks are: watermelon, shrimp, chicken, citrus fruits (when one of our ferrets would eat a sour grapefruit, he would become so excited he would give a little "laugh"), non-dairy whipped cream, tomatoes, cucumbers, raisins, dry breakfast cereal, crunchy treats, jello ...the list is virtually endless. You will have the fun of experimenting with your pet.

A word of caution about snacks: as much as we love to treat our ferrets, snacks should be only an occasional event. It's easy for a ferret to fill up on these foods and ignore the basic diet, which is what provides most of the nutrition.

Always provide your ferret with plenty of fresh, clean water.

A trio of ferrets cooling off on a summer day. Ferrets are extremely susceptible to heat stroke, so exercise caution when letting them out in hot weather.

A good diet can help to keep your ferret healthy and active throughout its life.

When selecting playthings for your ferret, choose only those that cannot be chewed into pieces.

FERRET PLAY

Ferrets have been compared to playful kittens, but unlike kittens, they do not lose their playfulness as they age. Although they require quite a bit of sleep, 12 to 14 hours a day, their play is very rambunctious and so much fun to share in and watch. The smallest items can provide endless games for ferrets—you're really only limited by your imagination. Members have told us of some of their pets' favorite toys and games, which are listed below: a large grocery sack to crawl into, the tube from paper towels, empty egg crate with string attached to use as a pull toy, dryer vent tubing, crinkly plastic bag (punch holes with a needle to prevent suffocation), ping pong balls, cat toys made of hard rubber, aluminum pie pan to pull around (makes satisfying noise on bare floor), dog squeaker toys (of hard rubber only). Lots of fun is to be found from half of a plastic Easter egg—the kind that separates in the middle. Put either half on a non-friction floor, flat side down, and watch an endless game of ferret hockey as your pet tries vainly to pick it up in his teeth/paws but succeeds only in pushing it ever farther.

Ferrets enjoy periods of rest. This ten-month-old female ferret uses a stuffed toy as her pillow.

Nap time for this little ferret. Try not to disturb your pet when it is sleeping: ferrets have been known to nip when they are abruptly awakened.

If you are going to provide tunnel-type objects, be sure they are large enough for your pet to fit through comfortably.

Some ferrets love to roughhouse. This ferret clasps its owner's hand, but it doesn't bite.

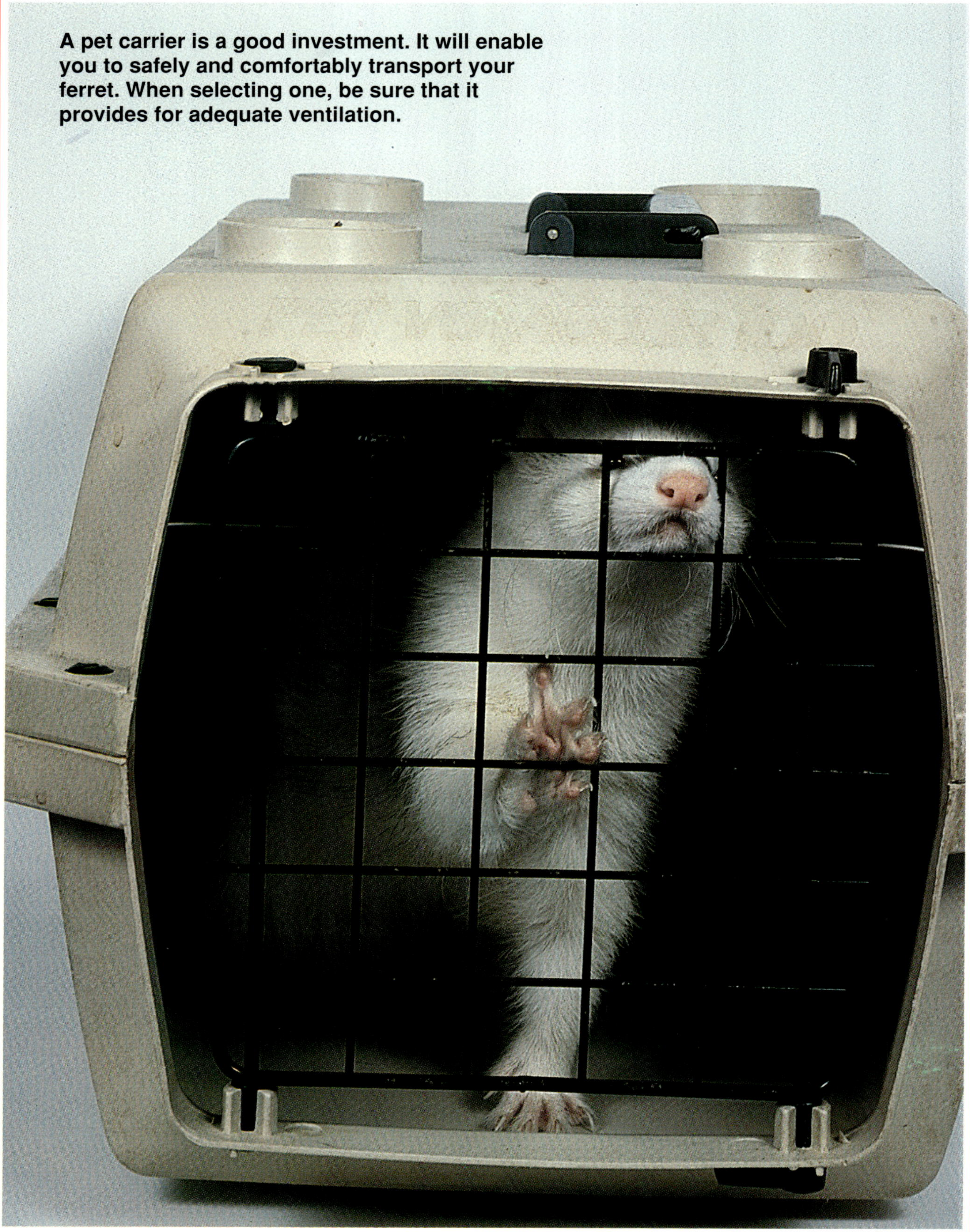

A pet carrier is a good investment. It will enable you to safely and comfortably transport your ferret. When selecting one, be sure that it provides for adequate ventilation.

TRAVEL

Ferrets are great travel companions and usually love to go with their owners on trips. When taking him/her in the car, use a small travel cage. One made for transporting cats will be a large temporary home for a ferret. Be sure it has a small litter pan and old towels in it. Cars are full of potential danger for an unconfined ferret. It's easy for a ferret to get under the driver's feet as the brake is applied, there are small areas under the dash where a ferret can escape, and it's easy for a little ferret to slip undetected from an open car door. Never leave the ferret in the car unattended on a warm day. If the temperature in the car goes over 75°, death could occur. Always be sure to have water available on even the shortest ride.

With proper planning, traveling with your ferret will not be a problem, and both of you will enjoy it.

If you have to leave your ferret home when you go on vacation, you will have to make arrangements to have a relative or friend regularly look in on him while you are away.

Many ferrets go on vacations with their owners. A little pre-planning makes the trip enjoyable for all concerned. There are still some states and cities which ban ferrets. But this problem can be solved by always traveling with your pet's latest medical records, including a note from the veterinarian stating that the ferret is owned by you and listing your present address. Naturally, all immunizations must be current.

Take the carrying cage—not only is it necessary for travel in the car but should also be used in homes of the persons you are visiting and in hotel/motel rooms. Since these places are not ferret proofed, your ferret could decide to explore its new surroundings and disappear.

One of our members had a story to relate regarding hotel exploration done by her ferret. It seems that she and her husband housed their ferret in its travel cage through the night. In the morning they let it run loose while they cleaned the cage. It got in the heating ducts of the hotel and appeared in the room of the guests on a lower floor. These people were not familiar with ferrets and, panic stricken, called the management to remove the "RODENT" from their room. Luckily, the owners and ferret were happily reunited.

Take along a supply of your ferret's food, as it might not be available in the area you are visiting, and travel time is not the time to be getting used to new food. Be sure water is always available.

Have your ferret wear a collar with its name and address tag at all times.

Mother and youngsters enjoying a play session. Adult ferrets retain the playful qualities exhibited by young ferrets.

You will find traveling with a ferret is a great "people attractor". This is a good time to do some positive PR for domestic ferrets. Let people know what great pets they are; however, it's better for the ferret's health not to be held or petted by strangers. Also, if the ferret becomes nervous and bites a stranger, in many towns the ferret can be seized and held for a rabies test (even though the rabies vaccine is current). At least one of the ways to test for rabies in a ferret is by decapitating the ferret. Don't let what started as a happy vacation end as a tragedy.

If you are taking public transportation—bus, train or plane, check with the carrier company well in advance of your trip. You will be advised about the best way to travel with a ferret. Some airlines treat pets in a carrier as luggage, forcing the pet to travel in the storage area of the plane. If this occurs, be certain this area is pressurized and temperature controlled. Other airlines will permit you to bring your carrying cage as a "carry-on bag", but only if there are no other pets on that particular flight. Policies vary not only from airline to airline, but seemingly from agent to agent. We suggest you start investigating this matter as soon as possible. Due to changes in pressure and altitude and the stress of flying, ferrets who are sick, frail or pregnant should not travel by plane. Flying can also be unsafe for older animals and those who are very nervous and high strung.

If you are planning international travel, check with the embassy of the country where you will visit. Ask if they have special rulings regarding ferrets, such as a mandatory quarantine period, and be sure to register your pet with U.S. Customs before leaving this country. A member who lives in Canada encountered problems. She informed us that to bring her ferret back from the United States, she went through a mountain of paperwork and expense.

Before you plan a trip into or out of Canada, call Agriculture Canada well in advance to avoid problems and expense. The numbers are (519) 837-9400 or (613) 952-8000.

Whenever you take a trip anywhere with your ferret, try to plan your stops. Many motels and campgrounds will accept small pets as guests. The FFC has a list of these facilities as well as a list of veterinarians throughout the United States who treat ferrets if the need arises.

If you are visiting a private home, it is definitely advisable to keep your ferret in a cage. Your host's home most likely is not ferret-proofed, and there are lots of hazards in unknown houses. A member reported, "I was visiting friends with my ferret, Wheezer. She was getting along well with my host's cat, so I decided to let her out of the small area I had sectioned off. I followed her around checking for holes in the cupboards and floors. Unfortunately, I took my eyes off her for only a few minutes and Wheezer disappeared. We tore everything apart—the cupboards, the floor—and then I found the HOLE. It went under the trailer down to the ground. I did everything the books suggested and more. I spent two long nights out in the cold calling and looking with no results."

LEASH TRAINING

Many ferrets are leash trained. Unfortunately, our ferrets never did catch on to this idea. To leash train your ferret, use a lightweight leash attached to your ferret's collar. Start by letting the ferret drag the leash around without you holding it. Then try walking the ferret in the house on the leash for no more than five minutes at a time. Be sure your pet can't wiggle out of the harness. If he/she seems to be used to it, you can try taking your pet outside—only a few minutes at a time to start.

Keep a close eye on your ferret. If he/she seems unduly frightened by traffic noise, footsteps of pedestrians or other animals, pick up your pet and bring it home. Some ferrets are pure homebodies and are best kept there. Never keep your pet unattended on a leash. Large dogs, cats or small children who think it is a toy could attack your ferret.

Certainly, if you live in an area where ferrets are banned or illegal, NEVER take them out in public.

If you walk your ferret on the sidewalk or in the streets, be aware that salt used to melt ice and slush is extremely painful on little paws. Don't walk your pet until all salt has disappeared—usually two to three weeks after the last salting. If salt gets into your ferret's paws, he/she will lick it out (after experiencing an extreme burning pain). The salt ingested is damaging to the system and can be fatal.

When traveling with your ferret, it should wear a collar and identification tag at all times. This will increase the chances of your pet's being returned should it get lost.

HEALTH PROBLEMS

Following is a list of health problems which may be encountered by ferrets. Be alert for symptoms; prompt veterinary attention can save lives and prevent suffering.

Anal Prolapse: This emergency condition is signaled by dragging the rear end and whimpering. It can be caused by the *E. coli* bacteria, which can be transmitted from ferret to humans and vice versa. It is treated by medication prescribed by the veterinarian.

Canine Distemper: This is always fatal in ferrets, and preventable by vaccine.

Cardiomyopathy: The heart muscle becomes thickened, causing it to pump less blood, leading to poor circulation. Symptoms include increased sleeping, collapsing during play, lethargy and poor appetite. More prevalent in males over three years old. Detected through a chest x-ray and controlled with medication.

Colds and Flu: Ferrets are susceptible to the same strain of cold and flu virus that affects humans. If someone in your home becomes ill with cold or flu, he should avoid handling or feeding the ferret. If this is not possible, carefully wash your hands before contact with your pet. Keep ferrets away from drafts and avoid sudden changes in temperature. Although it will be difficult to do, don't cuddle or kiss the ferret while you are ill. If your ferret develops symptoms of a cold, keep him/her warm, provide liquids and rest. Don't initiate or encourage active play. If symptoms last more than four days, consult your veterinarian.

Cushing's Disease: Symptoms of this disease include hair loss, frequent thirst, urination of large amounts, sluggishness, muscle loss and a potbellied appearance. Consult the veterinarian.

Diarrhea: If this persists for more than 24 hours and is accompanied by blood in the stool, or if the stool is black and tarry, if the ferret is depressed, lethargic or crying or has lost appetite, consult the veterinarian.

Ear Mites: These are tiny insects which, while not dangerous, will make your ferret miserable. They are usually found in the external ear canal. The first sign of infestation is the ferret frantically scratching his/her ears, shaking the head, crying or hitting the head against the wall. The ferret should be taken to the veterinarian immediately. The vet will remove a bit of the material from the ferret's ear and examine it microscopically. Once the diagnosis of ear mites has been made, medication will be administered. Be sure to follow instructions carefully and continue dosage as long as advised, even if the condition seems to have ended. Ear mites are contagious and can spread to other household

This ferret likes the flavor of a food supplement that the vet has recommended. Follow your vet's directions carefully when administering medications to your pet.

pets, so have all pets checked. Ear mites cannot be transmitted to humans.

Hair Loss: Natural hair loss occurs with the change in seasons. When ferrets live indoors with artificial light, shedding cycles can be disturbed. As long as the skin looks normal (not scabby or inflamed), there is probably no problem.

Hair Loss on Tail: This can be due to folliculitis, more prevalent in male ferrets. The pores on the tail become plugged-up blackheads. The veterinarian should be consulted and treatment started. Hair loss on the entire body, however, may be caused by as minor a problem as dry, forced air heat. A humidifier may be the answer.

Heartworm: This disease, transmitted by mosquitoes, is always fatal in ferrets. Even if your ferret never ventures outdoors, he/she is still susceptible. Consult your veterinarian about preventive measures.

Heat Exhaustion: Ferrets are extremely sensitive to temperatures above 80° F. When a ferret is overheated, it will lie on its stomach, breathing heavily in a shallow manner. Signs of heat exhaustion are: tongue hanging out, glazed eyes and inertia. The next stage is one from which the ferret may not recover—a coma. To prevent heat exhaustion, never leave your ferret in an unshaded area or keep him/her in a car (even with windows open) when temperature is above 72°. Be sure the cage is in an area with sufficient fresh air circulating around it, but avoid drafts. On extremely hot and humid days, limit your pet's exercise and play periods to times between 9 - 11 a.m. and after 4 p.m. Never permit your pet to be alone outside—always be near to watch for danger signs. At all times, be sure your ferret has a source of cool, fresh water.

A healthy ferret is an active ferret. A ferret owner should be familiar with his pet's normal behavior.

After your ferret has been outdoors, it should be checked for fleas and other parasites.

Ingestion of Foreign Objects: Ingestion of balloons, rubber bands, bits of sponges, pencil erasers, towels, cotton balls and latex rubber from soft rubber pet toys can cause intestinal blockage, which can be fatal. At the first sign of swelling, black, sticky or non-existent stools, vomiting, or finding evidence of chewed items, take your ferret to the veterinarian.

Ferrets are not timid animals. They are quick to investigate new things in their environment.

Insulinomas: Symptoms include seizures, glassy eyes, drooling and choking. Sugar applied to the gums, e.g., in the form of corn syrup, helps to bring them out of these seizures.

Lumps: Lumps under or on the skin should be examined by your veterinarian. If the lump is red and warm, it is probably an abscess, which should be treated immediately.

Lymphosarcoma: This is probably a viral disease, which can affect ferrets at any age. Because the disease weakens the immune system, the ferret may exhibit chronic illnesses such as digestive disorders, colds or flu. Lymph nodes may be enlarged. Lethargy, chronic diarrhea and rectal prolapse, loss of appetite, vomiting, weight loss and paralysis may occur. Consult your veterinarian.

Pancreatic Tumor: Symptoms include pawing at the mouth, excessive salivation, sluggish behavior, seizures and coma. Consult the veterinarian.

Pawing at the Mouth: This can be caused by gum or

Ferrets can get along with other family pets. Never leave a ferret and any other pet alone unless you are absolutely certain that they will not harm each other.

mouth inflammation or a pancreatic tumor. A careful examination by your veterinarian of the inner and outer gum surfaces should be done. Pawing can often indicate a feeling of nausea.

Skin Tumors: The most common type is mast cell tumor—a round, raised buttonlike bump. It is usually itchy. The other most common type is sebaceous gland adenoma, which has the appearance of an irregular growth and bleeds easily. Most skin tumors should be removed.

Stings: Those of a bee or wasp can be dangerous. Take your pet to a veterinarian at once.

Swallowing Poison: Act fast. Call your veterinarian or one of the emergency hotlines staffed by veterinarian toxicologists at the National Animal Poison Control Center (NAPCC). Be ready to give the name of the poison and/or to describe your pet's symptoms. Call NAPCC at 900/680-0000 or 800/548-2423 (have your VISA, MasterCard, Discovery or American Express card ready).

Valley Fever: This disease produces spores which become airborne and cause infection when inhaled. Symptoms are: a chronic respiratory infection or cough, skin lesion, joint enlargement, loss of weight and appetite, lethargy and fever. This disease is treatable, so consult your veterinarian if symptoms are noticed either singly or in a combination.

Vomiting—Withhold food for 6-8 hours. If vomiting continues, contains blood, or is associated with appetite loss or depression, consult the veterinarian.

A regular checkup for your ferret will include a physical exam, and, if they are due, booster shots.

Food bowls and water bottles should be cleaned regularly to prevent the build-up of harmful bacteria.

If your ferret shows an interest in a houseplant, he may be likely to use it as a litter pan.

GROOMING

To make your ferret socially acceptable, it will probably be necessary to bathe him/her about every ten days to two weeks (more often if they roll in dirt or get messy). I have heard many ferrets love baths but have never had this experience with any of mine.

BATHING

Bathing is best done in a kitchen sink which has been thoroughly washed. Put about six inches of tepid water in sink and gently lower the ferret into the water, talking softly and holding the body securely under the chest and hindquarters. Use tear-free baby shampoo or specially formulated ferret shampoo, taking care not to splash water into nose or face. Be sure, however, to thoroughly clean the face, as food can accumulate in the fur around the mouth and cause odor. Rinse thoroughly—two or three times to remove all soap.

Ferrets do not require any extensive grooming. Their short coats are very easy to maintain.

Ferrets will enjoy rolling in soft towels and will literally dry themselves. Most ferrets will turn into real acrobats, jumping and dancing from joy as they dry themselves. Be sure to avoid drafts. In cold weather, you may want to use a hair dryer (turned to lowest speed and cool setting) until the coat is thoroughly dry. If your ferret has dry skin or coat, your veterinarian can recommend the proper treatment.

There are products available that will eliminate a ferret's body odor and leave its coat soft and lustrous. Photo courtesy of Four Paws.

EAR CARE

Gently clean inside the ferret's ears with a cotton swab. You will probably find what you believe is dirt. This is normal—it is merely cerumen, or ear wax, which is secreted abundantly by ferrets. It is safe to leave it in the ear.

Ferrets should have their nails clipped regularly. Use a nail clipper especially designed for small animals.

NAIL CARE

Now is the time for a manicure. To distract the ferret, try putting a drop of a tasty supplement on the stomach. The ferret will be so busy licking it off, you can get the job done without fuss. Use a pet nail clipper, being careful not to cut too close to the bloodline (faintly pink area on nails).

Before and after bathing, it is important that your ferret not be exposed to drafts.

Bathing your ferret every ten days to two weeks will help to keep it odor free. In general, ferrets are not crazy about baths.

FLEAS

Most ferrets are affected by fleas—those in the northern states during warm months and those in the south on a constant basis. There are guidelines for flea control which can help immensely. Since ferrets are so small, it does not require a large number of fleas to cause anemia, and even death.

To check for flea infestation, place the ferret on a piece of white paper and comb through the fur from front to back. Look for any black spots that jump around or for egg clusters (little circular clumps.).

First, treat the ferret. There are several good products that are safe and effective. Some give an immediate flea-kill, but do not have long-term effects. They are safe enough to use several times a week. Flea combs work very well as ferrets usually learn to tolerate a few minutes of combing every day. A cat flea collar can be worn safely by a ferret, but be sure to expose the collar to air 24 hours before use.

Fleas are not an uncommon problem in a ferret. This is especially true if a ferret spends time outdoors.

Second, treat the environment. Wash or spray bedding. Vacuum your house thoroughly. Eggs and larvae inhabit the dust of vacuum cleaner bags. Remove all vacuum cleaner bags and replace with ones in which you have placed a teaspoon of flea powder, moth balls or flakes. This will continue to kill eggs, larvae and fleas that are vacuumed. Fog or spray the house using a product that is capable of killing the larval stage of the flea. If you have your house sprayed professionally, use caution. Some of the new granulated pesticides used by professionals can be highly toxic to ferrets, even though they carry no label warnings regarding dogs and cats.

Some ferrets are plagued with skin problems such as dry and flaky skin. There are products available that can help to provide relief from these conditions. Photo courtesy of Four Paws.

A ferret grooming itself. Skin problems are not always caused by parasites; sometimes other factors, such as diet, may be the cause.

If you are in doubt about the safety or efficacy of a flea product, call your veterinarian. As a general rule, those products that are safe for cats are also safe for ferrets.

Fleas also carry tapeworms, and ferrets are susceptible to them. If you notice small white worms in your ferret's stool, this is what they will be. This condition should be treated by a veterinarian.

However, not all scratching problems are due to fleas. If the scratching starts in the winter months, it is likely it is caused by dryness of the skin. This can be alleviated by raising the humidity in the room and adding a supplement to the diet. Check with your vet.

If you regularly let your pet outdoors, you should take precautions to prevent flea infestation. Check with your vet.

t.f.h.
THE WORLD'S BEST SMALL ANIMAL BOOKS
Established 1952
Established in 1952, T.F.H. Publications, Inc. is the world's largest publisher of books about pets.
C-138

SMALL ANIMALS

WORLD OF BATS
By Klaus Richarz & Alfred Limbrunner
TS-192
ISBN 0-86622-540-4
UPC 0-18214-25404-1
This visually stunning book contains information on all types of bats from all over the world. Many myths and legends involving bats are discussed and clarified scientifically.
HC, 7 3/4 x 10 1/2", 192 pages, over 150 full color photos.
$35.95

FERRETS
By Wendy Winsted, M.D.
KW-074
ISBN 0-86622-829-2
UPC 0-18214-28292-1
This book is designed for maximum information value and dramatic eye appeal. Topics covered include natural history, feeding, breeding, & health care.
HC, 5 1/2 x 8 1/2", 128 pages, full color photos throughout.
$9.95

A COMPLETE INTRODUCTION TO CHINCHILLAS
By Jack C. Harris
CO-042S
ISBN 0-86622-393-2
UPC 0-18214-23932-1
This highly informative book is perfect for beginners yet thorough enough for use by experienced owners.
SC, 5 1/2 x 8", 96 pages, over 60 full color photos & drawings.
$8.95

T.F.H. books are the most colorfully illustrated pet books in the world.

CAREERS FOR ANIMAL LOVERS
By Louise Miller
TS-168
ISBN 0-86622-169-7
UPC 0-1824 21697-1
Sound advice about which types of employment opportunities are available to people who want to work with animals, plus what type of training they may need to take advantage of those opportunities.
HC, 5 1/2 x 8 1/2", 176 pages.
$11.95

A STEP-BY-STEP BOOK ABOUT CHINCHILLAS
By Horst Kühner
SK-006
ISBN 0-86622-452-1
UPC 0-18214-24521-6
Humorous drawings and color photos combined with practical easy-to-read text make this book a great value for the hobbyist.
SC, 5 1/2 x 8 1/2", 64 pages, full color photos and drawings throughout.
$5.95

FERRETS AS A NEW PET
By Greg Ovechka
TU-014
ISBN 0-86622-622-2
UPC 0-18214-26222-0
The information in this book has been scientifically compiled to meet the needs and wants of new hobbyists. The design format makes reading a pleasure.
SC, 6 3/4 x 8 1/2", 64 pages, full color photos throughout.
$6.95

ALL ABOUT CHINCHILLAS
By Karen Zeinert
PS-845
ISBN 0-86622-143-3
UPC 0-18214-21433-5
Presents "soup to nuts" of Chinchilla care. Concise no-nonsense advice.
HC, 5 1/2 x 8 1/2", 128 pages, 49 full color photos.
$11.95

CHIPMUNKS
By Chris Henwood
KW-181
ISBN 0-86622-692-3
UPC 0-18214-26923-6
This book is designed for maximum information value and dramatic eye appeal. Topics covered include natural history, feeding, breeding, & health care.
HC, 5 1/2 x 8 1/2", 96 pages, full color photos and drawings.
$9.95

FERRETS AS A HOBBY
By Mary Field
TT-014
ISBN 0-86622-472-6
UPC 0-18214-24726-5
This popular book contains practical information to get you started as a hobbyist. The profits from all sales go toward preservation of Brazil's rain forest.
SC, 7 x 10", 96 pages, full color photos throughout.
$8.95

BREEDING CHINCHILLAS
By Egon Mösslacher
PS-850
ISBN 0-86622-804-7
UPC 0-18214-28047-7
Designed to be of special value to all of the new Chinchilla owners looking for good information.
HC, 5 1/2 x 8", 128 pages, 52 full color photos.
$17.95

THE PROPER CARE OF FANCY RATS
By Nick Mays
TW-122
ISBN 0-86622-340-1
UPC 0-18214-23401-2
Written with remarkable clarity, this book provides readers with all the basics, from selection to feeding, housing, & health care.
HC, 5 x 7", 256 pages, full color photos throughout.
$16.95

FERRETS IN YOUR HOME
By Wendy Winsted, M.D.
TS-106
ISBN 0-86622-988-4
UPC 0-18214-29884-7
The most comprehensive work on Ferrets detailing breeding and health care.
HC, 8 1/2 x 11", 160 pages, 90 full color photos & drawings.
$35.95

ALL ABOUT FERRETS
By Mervin F. Roberts
PS-754
ISBN 0-87666-914-3
UPC 0-18214-69143-3
A great introduction to Ferrets and how to take successful care of them.
SC, 5 1/2 x 8", 64 pages, full color photos throughout.
$4.95

A COMPLETE INTRODUCTION TO GERBILS
By Mrs. M. Ostrow
CO-018S
ISBN 0-86622-299-5
UPC 0-18214-22995-7
This highly informative book is perfect for beginners yet thorough enough for use by experienced owners.
SC, 5 1/2 x 8", 128 pages, over 92 full color photos & drawings.
$8.95

GERBILS
By D. G. Robinson, Jr.
PB-110
ISBN 0-86622-715-6
UPC 0-18214-27156-7
Expert practical advice for both beginners and those more advanced.
SC, 6 1/4 x 8 1/2", 80 pages, 32 full color photos.
$6.95

A STEP-BY-STEP BOOK ABOUT FERRETS
By Jay & Mary Field
SK-009
ISBN 0-86622-462-9
UPC 0-18214-24629-9
Humorous drawings and color photos combined with practical easy-to-read text make this book a great value for the hobbyist.
SC, 5 1/2 x 8 1/2", 64 pages, full color photos and drawings throughout.
$5.95

GERBILS AS A HOBBY
By Laurie Holland
TT-004
ISBN 0-86622-418-1
UPC 0-18214-24181-2
This popular book contains practical information to get you started as a hobbyist. The profits from all sales go toward preservation of Brazil's rain forest.
SC, 7 x 10", 96 pages, full color photos throughout.
$8.95

THE PROFESSIONAL'S BOOK OF GERBILS
By Robert Bernhard
TS-120
ISBN 0-86622-669-9
UPC 0-18214-26699-0
A highly informative book covering all aspects of keeping and breeding Gerbils.
HC, 8 1/2 x 11", 160 pages, over 100 full color photos.
$17.95

YOUR FIRST FERRET
By Adelle Porch
YF-105
ISBN 0-86622-115-8
UPC 0-18214-21158-7
Designed specifically for use by beginners, this colorful book concentrates on providing just the right amount of information.
SC, 6 x 7 1/2", 32 pages, full color photos throughout.
$2.29

GERBILS AS A NEW PET
By Anmarie Barrie
TU-002
ISBN 0-86622-608-7
UPC 0-18214-26087-5
The information in this book has been scientifically compiled to meet the needs and wants of new hobbyists. The design format makes reading a pleasure.
SC, 6 3/4 x 8 1/2", 64 pages, full color photos throughout.
$6.95

THE PROPER CARE OF GERBILS
By Tony David
TW-106
ISBN 0-86622-187-5
UPC 0-18214-21875-3
Written with remarkable clarity, this book provides readers with all the basics, from selection to feeding, housing, & health care.
HC, 5 x 7", 256 pages, full color photos throughout.
$16.95

T.F.H. books are the most colorfully illustrated pet books in the world.

GERBILS
By Paul Paradise
KW-037
ISBN 0-87666-927-5
UPC 0-18214-69275-1
This book is designed for maximum information value and dramatic eye appeal. Topics covered include natural history, feeding, breeding, & health care.
HC, 5 1/2 x 8 1/2", 96 pages, full color photos and drawings.
$9.95

A STEP-BY-STEP BOOK ABOUT GERBILS
By P. Bradley & H. Pence
SK-011
ISBN 0-86622-467-X
UPC 0-18214-80060-6
Humorous drawings and color photos combined with practical easy-to-read text make this book a great value for the hobbyist.
SC, 5 1/2 x 8 1/2", 64 pages, full color photos and drawings throughout.
$5.95

YOUR FIRST GERBIL
By Leroy Mervan
YF-107
ISBN 0-86622-063-1
UPC 0-18214-20631-6
Designed specifically for use by beginners, this colorful book concentrate on providing just the right amount of information.
SC, 6 x 7 1/2", 32 pages, full color photos throughout.
$2.29

GUINEA PIGS FOR THOSE WHO CARE
By Dennis Kelsey-Wood
B-104
ISBN 0-7938-1385-9
UPC 0-18214-11385-0
Easy-to-read text combined with stunning graphic design makes this book a must for the beginning enthusiast.
SC, 7 x 8 1/2", 32 pages, full color photos throughout.
$4.95

GUINEA PIGS AS A NEW PET
By Stephen Nelson
TU-006
ISBN 0-86622-613-3
UPC 0-18214-26133-9
The information in this book has been scientifically compiled to meet the needs and wants of new hobbyists. The design format makes reading a pleasure.
SC, 6 3/4 x 8 1/2", 64 pages, full color photos throughout.
$6.95

GUINEA PIGS
By Kay Ragland
KW-016
ISBN 0-86622-830-6
UPC 0-18214-28306-5
This book is designed for maximum information value and dramatic eye appeal. Topics covered include natural history, feeding, breeding, & health care.
HC, 5 1/2 x 8 1/2", 128 pages, full color photos and drawings.
$9.95

PROPER CARE OF GUINEA PIGS
By Peter Gurney
TW-108
ISBN 0-86622-195-6
UPC 0-18214-21956-9
Written with remarkable clarity, this book provides readers with all the basics, from selection to feeding, housing, & health care.
HC, 5 x 7", 256 pages, full color photos throughout.
$16.95

A BEGINNER'S GUIDE TO GUINEA PIGS
By Tom Willkie
T-105
ISBN 0-86622-304-5
UPC 0-18214-23045-8
Great reference source for those just getting started.
HC, 5 1/2 x 8 1/2", 64 pages, full color photos throughout.
$4.95

GUINEA PIGS
By Carolyn Ruf
PB-113
ISBN 0-86622-987-6
UPC 0-18214-29876-2
Easy-to-read text provides practical advice for both beginners and those more advanced.
SC, 6 1/4 x 8 1/2", 80 pages, full color photos throughout.
$6.95

A STEP-BY-STEP BOOK ABOUT GUINEA PIGS
By Anmarie Barrie
SK-013
ISBN 0-86622-450-5
UPC 0-18214-24505-6
Humorous drawings and color photos combined with practical easy-to-read text make this book a great value for the hobbyist.
SC, 5 1/2 x 8 1/2", 64 pages, full color photos and drawings throughout.
$5.95

T.F.H. offers the world's widest selection of books dealing with small animals of every type.

GUINEA PIGS AS A HOBBY
By Anmarie Barrie
TT-008
ISBN 0-86622-420-3
UPC 0-18214-24203-1
This popular book contains practical information to get you started as a hobbyist. The profits from all sales go toward preservation of Brazil's rain forest.
SC, 7 x 10", 96 pages, full color photos throughout.
$8.95

A COMPLETE INTRODUCTION TO GUINEA PIGS
By Margaret Elward
CO-038S
ISBN 0-86622-383-5
UPC 0-18214-23835-5
This highly informative book is perfect for beginners yet thorough enough for use by experienced owners.
SC, 5 1/2 x 8", 96 pages, over 92 full color photos & drawings.
$8.95

YOUR FIRST GUINEA PIG
By Anja J. Steinkamp
YF-109
ISBN 0-86622-066-6
UPC 0-18214-20666-8
Designed specifically for use by beginners, this colorful book concentrates on providing just the right amount of information.
SC, 6 x 7 1/2", 32 pages, full color photos throughout.
$2.29

HAMSTERS
By Percy Parslow
KW-015
ISBN 0-86622-28314
UPC 0-18214-28314-0
This book is designed for maximum information value and dramatic eye appeal. Topics covered include natural history, feeding, breeding, & health care.
HC, 5 1/2 x 8 1/2", 128 pages, full color photos and drawings.
$9.95

A STEP-BY-STEP BOOK ABOUT DWARF HAMSTERS
By Anmarie Barrie
SK-040
ISBN 0-86622-479-3
UPC 0-18214-24793-7
Humorous drawings and color photos combined with practical easy-to-read text make this book a great value for the hobbyist.
SC, 5 1/2 x 8 1/2", 64 pages, full color photos and drawings throughout.
$5.95

A BEGINNERS GUIDE TO HAMSTERS
By Jay Stephens
T-106
ISBN 0-86622-305-3
UPC 0-18214-23053-3
Great reference source for those just getting started.
HC, 5 1/2 x 8 1/2", 64 pages, full color photos throughout.
$4.95

HAMSTERS
By Dr. G. Edgar Folk, Jr.
PB-114
ISBN 0-86622-802-0
UPC 0-18214-28020-0
Easy-to-read text provides practical advice for beginners as well as those more advanced.
SC, 6 1/4 x 8 1/2", 80 pages, full color photos throughout.
$6.95

HAMSTERS FOR THOSE WHO CARE
By Anmarie Barrie
B-105
ISBN 0-7938-1376-X
UPC 0-18214-11376-0
Easy-to-read text combined with stunning graphic design make this book a must for the beginning enthusiast.
SC, 7 x 8 1/2", 32 pages, full color photos throughout.
$4.95

BREEDING HAMSTERS
By Marshall Ostrow
KW-134
ISBN 0-86622-564-1
UPC 0-18214-25641-0
This book is designed for maximum information value and dramatic eye appeal. Topics covered include feeding & health care with strong emphasis on breeding techniques.
HC, 5 1/2 x 8 1/2", 96 pages, full color photos and drawings.
$9.95

HAMSTERS AS A HOBBY
By Grey Ovechka
TT-006
ISBN 0-86622-412-2
UPC 0-18214-24122-5
This popular book contains practical information to get you started as a hobbyist. The profits from all sales go toward preservation of Brazil's rain forest.
SC, 7 x 10", 96 pages, full color photos throughout.
$8.95

PROPER CARE OF HAMSTERS
By Marianne Mays
TW-125
ISBN 0-86622-370-3
UPC 0-18214-23703-7
Written with remarkable clarity, this book provides readers with all the basics, from selection to feeding, housing, & health care.
HC, 5 x 7", 256 pages, full color photos throughout.
$16.95

HAMSTERS AS A NEW PET
By Anmarie Barrie
TU-003
ISBN 0-86622-610-9
UPC 0-18214-26109-4
The information in this book has been scientifically compiled to meet the needs and wants of new hobbyists. The design format makes reading a pleasure.
SC, 6 3/4 x 8 1/2", 64 pages, full color photos throughout.
$6.95

A COMPLETE INTRODUCTION TO HAMSTERS
By Mervin Roberts
CO-020S
ISBN 0-86622-282-0
UPC 0-18214-22820-2
This highly informative book is perfect for beginners yet thorough enough for use by experienced owners.
SC, 5 1/2 x 8", 96 pages, over 92 full color photos & drawings.
$8.95

A STEP-BY-STEP BOOK ABOUT HAMSTERS
By Anmarie Barrie
SK-014
ISBN 0-86622-458-0
UPC 0-18214-24580-3
Humorous drawings and color photos combined with practical easy-to-read text make this book a great value for the hobbyist.
SC, 5 1/2 x 8 1/2", 64 pages, full color photos and drawings throughout.
$5.95

YOUR FIRST HAMSTER
By Peter Smith
YF-110
ISBN 0-86622-067-4
UPC 0-18214-20674-3
Designed specifically for use by beginners, this colorful book concentrates on providing just the right amount of information.
SC, 6 x 7 1/2", 32 pages, full color photos throughout.
$2.29

POT-BELLIED PIGS AND OTHER MINIATURE PET PIGS
By Lisa Hall Huckaby
TS-181
ISBN 0-86622-438-6
UPC 0-18214-24386-1
Learn the "real" facts about pot-bellies while enjoying the full color candor of this porcine spectacular.
HC, 5 1/2 x 8 1/2", 144 pages full color photos throughout.
$23.95

MICE AS A NEW PET
By Richard Pfarr
TU-022
ISBN 0-86622-530-7
UPC 0-18214-25307-5
The information in this book has been scientifically compiled to meet the needs and wants of new hobbyists. The design format makes reading a pleasure.
SC, 6 3/4 x 8 1/2", 64 pages, full color photos throughout.
$6.95

AFRICAN PYGMY HEDGEHOGS
By Dennis Kelsey-Wood
TS-238
ISBN 0-7938-0192-3
UPC 0-18214-10192-5
This book is specifically designed to provide beginners with information wanted and needed most.
HC, 7 x 8 1/2", 160 pages full color photos throughout.
$17.95

POT-BELLIED PIGS AS A FAMILY PET
By Michael Taylor
TS-188
ISBN 0-86622-081-X
UPC 0-18214-80036-1
This book covers important facts about the pig family including purchasing, housing, caring for, feeding, and training.
HC, 7 x 10", 192 pages, full color photos throughout.
$35.95

FANCY MICE
By Chris Henwood
KW-224
ISBN 0-96622-342-8
UPC 0-18214-23428-9
This book is designed for maximum information value and dramatic eye appeal. Topics covered include feeding, breeding, & health care.
HC, 5 1/2 x 8 1/2", 96 pages, full color photos and drawings.
$9.95

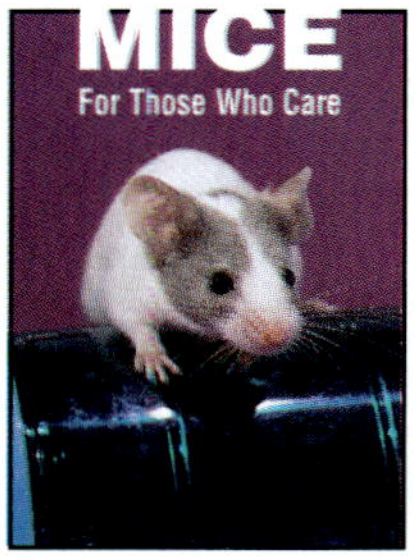

MICE FOR THOSE WHO CARE
By Dennis Kelsey-Wood
B-110
ISBN 0-7938-1388-3
UPC 0-18214-11388-1
Easy-to-read text combined with stunning graphic design make this book a must for the beginning enthusiast.
SC, 7 x 8 1/2", 32 pages, full color photos throughout.
$4.95

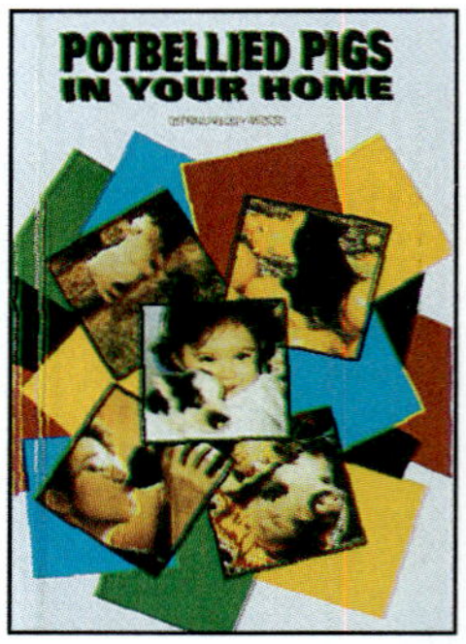

POT-BELLIED PIGS IN YOUR HOME
By Dennis Kelsey-Wood
TS-208
ISBN 0-7938-0075-7
UPC 0-18214-10075-1
Explore the special personalities and needs of pot-bellies with all the information you need about caring for them.
HC, 7 x 10", 160 pages, full color photos throughout.
$35.95

T.F.H. books are the most colorfully illustrated pet books in the world.

MICE AS A HOBBY
By Jack Young
TT-019
ISBN 0-86622-949-3
UPC 0-18214-29493-1
This popular book contains practical information to get you started as a hobbyist. The profits from all sales go toward preservation of Brazil's rain forest.
SC, 7 x 10", 96 pages, full color photos throughout.
$8.95

YOUR FIRST MOUSE
By Nick Mays
YF-120
ISBN 0-7938-0179-6
UPC 0-18214-10179-6
Designed specifically for use by beginners, this colorful book concentrates on providing just the right amount of information.
SC, 6 x 7 1/2", 32 pages, full color photos throughout.
$2.29

RABBITS
By Paul Paradise
KW-021
ISBN 0-86622-832-2
UPC 0-18214-28322-5
This book is designed for maximum information value and dramatic eye appeal. Topics covered include natural history, feeding, breeding, & health care.
HC, 5 1/2 x 8 1/2", 96 pages, full color photos and drawings.
$9.95

RABBITS
By Mervin F. Roberts
PB-124
ISBN 0-86622-844-6
UPC 0-18214-28446-8
Easy-to-read text provides practical advice for beginners as well as those more advanced.
SC, 6 1/4 x 8 1/2", 80 pages, full color photos throughout.
$6.95

DWARF RABBITS AS A NEW PET
By Andrea Dieker & Jutta Steinkamp
TU-026
ISBN 0-86622-537-4
UPC 0-18214-25374-7
The information in this book has been scientifically compiled to meet the needs and wants of new hobbyists. The design format makes reading a pleasure.
SC, 6 3/4 x 8 1/2", 64 pages, full color photos throughout.
$6.95

A COMPLETE INTRODUCTION TO RABBITS
By Marshal Merton
CO-022S
ISBN 0-86622-281-2
UPC 0-18214-22812-7
This highly informative book is perfect for beginners yet thorough enough for use by experienced owners.
SC, 5 1/2 x 8", 96 pages, over 96 full color photos & drawings.
$8.95

RABBITS AS A HOBBY
By Bob Bennett
TT-003
ISBN 0-86622-417-3
UPC 0-18214-24173-7
This popular book contains practical information to get you started as a hobbyist. The profits from all sales go toward preservation of Brazil's rain forest.
SC, 7 x 10", 96 pages, full color photos throughout.
$8.95

PROPER CARE OF DWARF RABBITS
By Michael Mettler
TW-121
ISBN 0-86622-443-2
UPC 0-18214-24432-5
Written with remarkable clarity, this book provides readers with all the basics, from selection to feeding, housing, & health care.
HC, 5 x 7", 256 pages, full color photos throughout.
$16.95

DWARF RABBITS
By Gunter Flauaus
H-1073
ISBN 0-86622-671-0
UPC 0-18214-26710-2
An introductory text for new owners and potential owners.
HC, 5 1/2 x 8", 128 pages, B & W and full color photos.
$17.95

RABBITS AS A NEW PET
By Barry Martin
TU-010
ISBN 0-86622-618-4
UPC 0-18214-26184-1
The information in this book has been scientifically compiled to meet the needs and wants of new hobbyists. The design format makes reading a pleasure.
SC, 6 3/4 x 8 1/2", 64 pages, full color photos throughout.
$6.95

ENCYCLOPEDIA OF RABBITS
By D. Robinson
H-984
ISBN 0-87666-911-9
UPC 0-18214-69119-8
An interesting and useful book blending practical advice about maintenance & breeding with a large assemblage of full color photos.
HC, 5 1/2 x 8", 320 pages, B & W and full color photos.
$19.95

T.F.H. offers the world's widest selection of books dealing with small animals of every type.

ALL ABOUT RABBITS
By Howard Hirschhorn
M-543
ISBN 0-86622-693-1
UPC 0-18214-26931-1
An excellent book for someone buying their first rabbit that is interested in breeding and showing.
HC, 5 1/2 x 8", 96 pages, B & W and full color photos.
$11.95

DWARF RABBITS AS A HOBBY
By Dennis Kelsey-Wood
TT-023
ISBN 0-86622-713-X
UPC 0-18214-80061-3
This popular book contains practical information to get you started as a hobbyist. The profits from all sales go toward preservation of Brazil's rain forest.
SC, 7 x 10", 96 pages, full color photos throughout.
$8.95

RABBITS FOR THOSE WHO CARE
By Anmarie Barrie
B-117
ISBN 0-7938-1377-8
UPC 0-18214-11377-5
Easy-to-read text combined with stunning graphic design make this book a must for the beginning enthusiast.
SC, 7 x 8 1/2", 32 pages, full color photos throughout.
$4.95

RABBITS, LOOK & LEARN
By Marcy Myerovich
KD-004S
ISBN 0-7938-0171-0
UPC 0-18214-10171-0
This exciting book contains valuable and authoritative information combined with beautifully silhouetted photos.
SC, 6 3/4 x 10", 64 pages, full color photos throughout.
$6.95

STARTING RIGHT WITH RABBITS
By Mervin F. Roberts
PS-796
ISBN 0-87666-814-7
UPC 0-18214-68147-2
Contains a wealth of information about how to handle your pet rabbit.
HC, 8 1/2 x 11", 112 pages, over 50 full color photos.
$11.95

RATS AS A HOBBY
By Susan Fox
TT-031
ISBN
UPC 0-18214-12182-4
This popular book contains practical information to get you started as a hobbyist. The profits from all sales go toward preservation of Brazil's rain forest.
SC, 7 x 10", 96 pages, full color photos throughout.
$8.95

LOP RABBITS
By Sandy Crook
PS-809
ISBN 0-86622-137-9
UPC 0-18214-21379-6
This book provides a rare learning experience for anyone interested in these pets.
HC, 5 1/2 x 8", 128 pages, B & W and full color photos.
$24.95

A STEP-BY-STEP BOOK ABOUT RABBITS
By Anmarie Barrie
SK-001
ISBN 0-86622-475-0
UPC 0-18214-24750-0
Humorous drawings and color photos combined with practical easy-to-read text make this book a great value for the hobbyist.
SC, 5 1/2 x 8 1/2", 64 pages, full color photos and drawings throughout.
$5.95

RATS FOR THOSE WHO CARE
By Dennis Kelsey-Wood
B-114
ISBN 0-7938-1392-1
UPC 0-18214-11392-8
Easy-to-read text combined with stunning graphic design makes this book a must for the beginning enthusiast.
SC, 7 x 8 1/2", 32 pages, full color photos throughout.
$4.95

PROPER CARE OF RABBITS
By D. Camphell
TW-110
ISBN 0-86622-196-4
UPC 0-18214-21964-4
Written with remarkable clarity, this book provides readers with all the basics, from selection to feeding, housing, & health care.
HC, 5 x 7", 256 pages, full color photos throughout.
$16.95

YOUR FIRST RABBIT
By Louise Vernier
YF-114
ISBN 0-86622-071-2
UPC 0-18214-20712-2
Designed specifically for use by beginners, this colorful book concentrates on providing just the right amount of information.
SC, 6 x 7 1/2", 32 pages, full color photos throughout.
$2.29

STICK INSECTS
By David Alderton
SK-043
ISBN 0-86622-349-5
UPC 0-18214-23495-1
Humorous drawings and color photos combined with practical easy-to-read text make this book a great value for the hobbyist.
SC, 5 1/2 x 8 1/2", 64 pages, full color photos and drawings throughout.
$5.95

RATS
By Susan Fox
KW-128
ISBN 0-87666-933-X
UPC 0-18214-80004-0
This book is designed for maximum information value and dramatic eye appeal. Topics covered include feeding, breeding, & health care.
HC, 5 1/2 x 8 1/2", 96 pages, full color photos and drawings.
$9.95

TFH PUBLICATIONS, INC.
One TFH Plaza • Neptune, NJ 07753
U.S.A.

TFH PUBLICATIONS / NYLABONE LTD
P.O. Box 15 • Waterlooville, PO76BQ
England

Regular brushing is not necessary. Additionally, most ferrets won't tolerate it.

BREEDING

Breeding of ferrets is a matter to be considered seriously. It is not easy for a novice—many things can go wrong. The jill may be physically unable to bear kits, or she may prove to be a bad mother, refusing to nurse the kits or even attacking and killing them. If she is unable to nurse her kits, they will starve to death unless a surrogate mother can be found or the kits are fed a special formula.

Another problem you will face is the "adoption" of the kits. Can you be sure they will be placed in loving homes or will they be acquired as a novelty and ignored once the newness has worn off? Also, prospective owners should be informed about upkeep costs (physical exams, vaccines, etc.). In most cases, if the new owners are unwilling or unable to spend the amount required on medical care, the ferret will suffer needlessly.

The FFC strongly urges neutering for all ferrets unless they are owned by professional breeders. Neutering also makes males more docile and both sexes sweeter smelling.

The first estrus (heat) usually occurs at four to five months of age and will occur twice a year thereafter. Estrus causes a change in the appearance of the vulva, which becomes swollen. At the height of estrus, there might be a slight clear discharge.

A jill in heat must be brought out of heat or bred. If neither of these happens, she will develop aplastic anemia, which is fatal. A safe alternative to end estrus is an injection administered by the veterinarian followed by spaying.

Pet-quality ferrets must be raised with abundant human social interaction from the moment they open their eyes (usually the 27th day after birth). Playing, cuddling and touching ferrets (which comes naturally to most of us) is essential if ferrets are to become good pets. Ferrets relate to their litter mates by vigorously wrestling and biting each other. They have very strong skin around the head, neck and shoulders, which is where they grab each other and hang on. They are not angry or aggressive when they do this—just curious, fun-loving ferrets. If they play like this during most of their waking hours, how can we expect them to react to human touch in a different manner unless they are socialized?

Ferrets mating. The hob bites the scruff of the jill's neck until she submits to him.

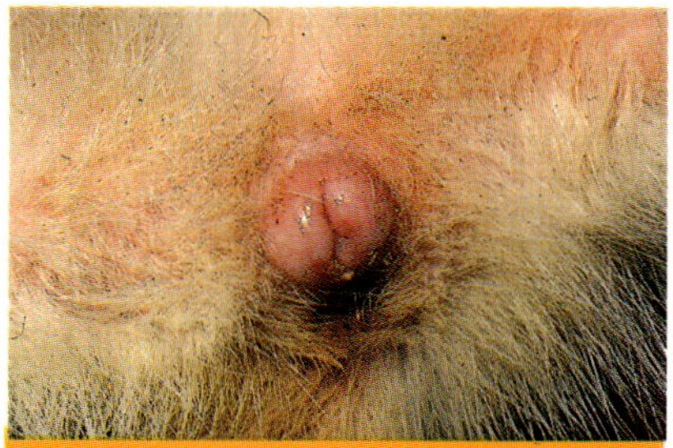

The vulva appears noticeably different when a female is coming into heat.

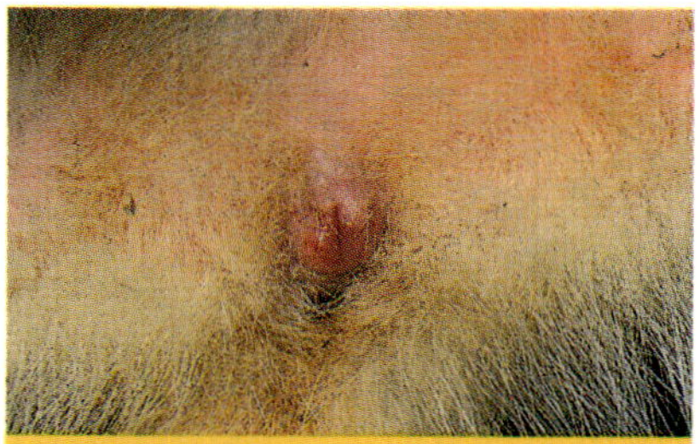

Genitalia of a female ferret in non-breeding condition.

This young ferret is being carried back to the nest by its mother. It is 23 days old.

An average-size litter of ferrets. At birth, ferrets are blind, deaf, and hairless.

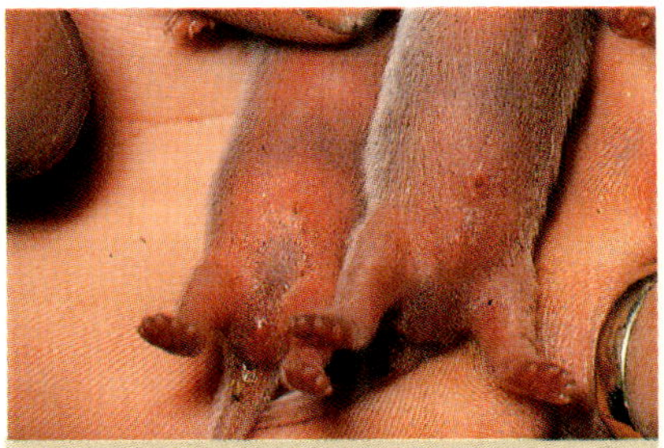

Two-day-old babies. Breeding entails many responsibilities. Unless you have the necessary time and resources, you should not undertake such a venture.

These babies are one week old. Their fur, which has the appearance of whitish fuzz, will not be completely grown in for several more weeks.

If after considering the above points you still decide to breed ferrets, we offer the following breeding tips.

The most fertile time for a jill is the tenth day of the heat cycle. At that time place the hob and jill together. The breeding process is a violent one—the hob will bite the back of the jill's neck and drag her around the cage until she submits. The jill will usually scream and fight. Although this looks frightening, it is typical ferret "courtship". Remove the male from the cage after breeding.

Gestation period is usually 42 days. If your jill's diet is adequate, no dietary change is necessary.

Prior to the birth, isolate the expectant mother from other ferrets. Put her in a nestbox (a plastic dishpan is good for this purpose), lined with a small towel, placed in a quiet room. The jill will make her own nest. The nest must be kept very clean at all times.

Generally the jill will get through the birthing process on her own in about two hours, biting the umbilical cords, ingesting the afterbirth, and cleaning the kits. However, if possible, it is best to be on hand if assistance or emergency veterinary care is needed.

The kits are hairless, pinkish red and blind at birth. Check regularly on the litter for several days to be sure none of the kits is rejected by the mother. If rejection occurs, the kit will have to be fed by a surrogate mother or hand fed with a milk replacement product.

The eyes usually open on the twenty-seventh day. From that time on, the kits must be handled gently by humans in order to become pet stock ferrets.

At the age of four weeks, weaning can be started by teaching the kits to drink from a dish and providing water-soaked ferret chow. Permanent canine teeth grow in when the kit is seven weeks old, and at that time they can handle food that is soaked very little.

After weaning has been completed, ferrets can be taken from their mother and be placed for adoption.

Baby ferrets must be handled very gently.

This kit is just over two weeks old. It is still too early to determine what its color and markings will be as an adult.

THE SENIOR FERRET

The aging process with ferrets is not as marked as that of dogs or cats. Healthy senior ferrets stay playful and energetic, in many cases, through middle and old age. However, there are a few changes owners will notice.

After the age of five or six, their coats may not be as full or luxuriant in the winter. As long as there are no bald spots or heavy shedding, this is nothing to be concerned about.

The older ferret will increase his/her sleeping time as the aging process goes on. This is normal—let your pet set the pace. The ferret will still enjoy playtime with you—he or she will let you know when it's time to play. The ferret will continue to thrive on love and companionship, and a nap or rest on a lap is a great treat to both owner and ferret.

Silver-mitt female.

Keep an eye on your pet's food dish—note any drop in appetite and report this to your veterinarian. This is especially important if more than one ferret shares the food dish; the food level may go down, but one ferret may not be eating. If the ferret has developed problems with his/her teeth or gums, this will also interfere with the ferret receiving necessary nutrients.

Note your ferret's stool. If there is a change from normal, contact your veterinarian.

The pads of the feet may become hard and dry. A dab of Vaseline daily will improve the condition.

If possible, keep your pet's bed, food dish, toys, etc. in the same place. Ferrets are creatures of habit, and routine is comforting.

Deborah Kemmerer, DVM, says that conditions which might develop in the senior ferret are: cataracts, dental and gum disease, dull and thinning coats, arthritis, chronic kidney failure, and increased incidence of tumors may occur in ferrets over the age of five. Cataracts are not really harmful in themselves; most ferrets do very well even with severe cataracts. If they form rapidly, however, the ferret should be examined to determine if the cause is not due to diabetes or Cushing's Disease.

A geriatric workup should be done annually. It should include a blood count and whole body x-ray.

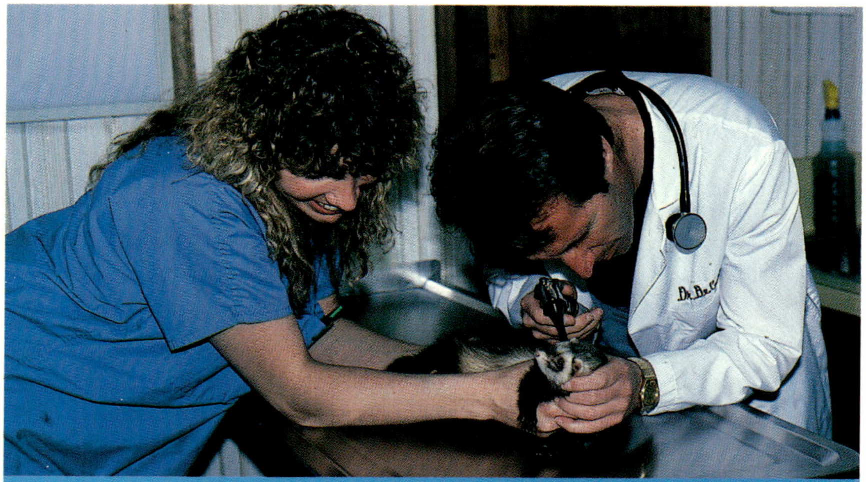

An annual check-up is a must for the senior ferret. Be sure to report to the vet any changes in your pet's appearance or behavior.

Older ferrets frequently develop gum disease and lose teeth due to severe tartar formation. The older ferret's teeth should be cleaned professionally every one or two years. To help prevent tooth and gum disease, tartar can be cleaned from the teeth when the ferret is sedated.

Kidney failure can be treated in part by a change in diet to a lower protein food. Prescription diets may be necessary.

As ferrets become older, the hair may become drier. Bathe as seldom as possible, and then use a gentle shampoo.

The most serious problem for older ferrets is an increase in tumors. The most common ones in older ferrets are insulinoma, lymphosarcoma, and adrenal tumors, all of which can be treated.

Adhere to the schedule of examinations and vaccinations prescribed by your veterinarian. If your pet is on medication, be sure to administer it exactly as ordered.

EUTHANASIA

If the time comes when all that is medically possible has been done for your pet and he/she no longer seems to be enjoying life, or is suffering unavoidable pain, it will be necessary to consider euthanasia. This, although very difficult for owners, is the final gift of love to offer your pet. A few months after my husband passed away, I was faced with this problem. Little Furry was too ill to be helped by veterinary care, and it would have been cruel to prolong his life. After feeding Furry all the raisins he wanted (very few at that point), I asked the veterinarian to administer the injection which would relieve his suffering. I held him in my arms until he fell asleep.

Dr. Jerry Rousseau, a psychotherapist in Milwaukee, Wisconsin, spoke about this subject. He observed, "When you open up your heart to allow yourself to be close to a ferret, you also open yourself up to feeling a sense of loss when that ferret dies."

Dr. Rosseau said that as one confronts the death of a ferret, he/she will experience a number of different stages of grief:

1. The first stage is shock, disbelief. There is a feeling of numbness that is important, for it will help make the initial adjustment to the loss.

2. The second stage has to do with anger. Sometimes we try to blame ourselves for the ferret's death. "If I only..." Do

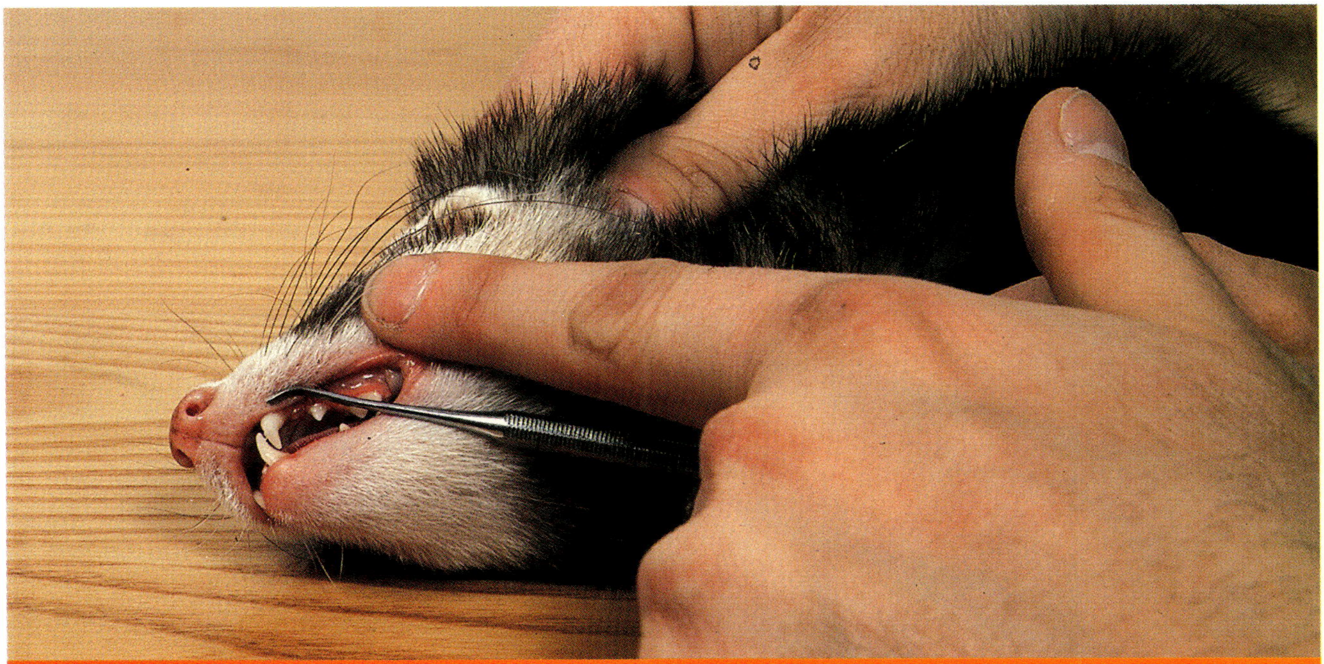

The older ferret will need regular attention to its teeth to help prevent tooth loss and gum disease.

The changes that occur with aging are not as noticeable in ferrets as they are in other pets.

not try to push angry feelings away—they are natural. Accept them to their fullest; they will eventually run their course.

3. The third stage has to do with sadness—the things we are going to miss. Let yourself experience the loss. Find people you feel close with to share your feelings. Talk to other ferret owners. Don't keep it to yourself, share it.

4. The fourth stage has to do with acceptance. "I'm sorry and I'm sad this has happened, but it really has happened in my life."

5. The fifth stage is moving on: not forgetting, running away, or suppressing your feelings, but being able to move on in your life. It is good therapy to sit down and experience each stage consciously.

Some ferrets may become less active as they age, but they will still be interested in their surroundings.

Sable female ferret.

THE FERRET FANCIERS CLUB

We cordially invite you to join our more than 3,000 members around the world. The FFC (a non-profit organization) offers a bi-monthly newsletter which gives updates of medical information, ferret products and, perhaps most importantly, an exchange of ferret questions and information. Members write in with problems or questions, and we are deluged with answers from members who have encountered (and usually solved) the problem. Additionally, we offer a veterinary consulting service; if your veterinarian is having a problem with a diagnosis or treatment, we can put him/her in contact with a member veterinarian who has experience in that particular area. We offer support and advice for ferret owners trying to legalize ferrets in areas in which they are presently banned and provide literature about a vast number of ferret-related concerns.

The FFC also provides a registry service. This enables us to have a more accurate idea of the number of pet ferrets when we attempt to obtain more favorable legislation regarding our pets. All ferrets can be registered. To register a ferret, send the following information to: FFC Registration, 711 Chautauqua Court, Pittsburgh, PA 15214. List owner's name, address, ferret's date of birth (if known), sex, color, and name desired. Also list second and third choice of names in the event of duplication. If sire and dam are known, also send this information. After a search of our records, a card will be sent to you which will be your official registration. The cost is $10 per ferret (US currency).

For information on joining the club write: FFC, 711 Chautauqua Court, Pittsburgh, PA 15214 or call (412) 322-1161. This number also is a 24-hour hot line for ferret emergencies.

May I take this opportunity to wish you every happiness with your new pet, the fabulous ferret.

A litter of 37-day-old ferrets. Joining a ferret club is an excellent way to keep up with the newest developments in keeping these pets.

Ferrets are natural acrobats!

Don't bathe your ferret too frequently. Otherwise, you will strip the coat of essential oils, and skin problems such as dryness and itchiness may develop.

FERRET OWNERS' QUESTIONS

In the ten years the FFC has been in existence, we have received many questions from members. We would like to address some of those most frequently asked.

Q. I have only one ferret. Is it a good idea to get another?

A. Two ferrets will keep each other company when you are away. They will spend much of their pent-up energy playing together and will always have each other to cuddle up with. However, another ferret will not substitute for human companionship—ferrets thrive on it. Before you rush out to get a second pet, however, consider whether you will have enough time to care for another ferret, have a cage large enough for two, be able to afford vaccine, food and medical care for another pet, and be willing to give the time and patience to introduce the pets to each other.

Q. If I decide to get another ferret, how will I introduce the new pet to the one I already have?

A. This can be tricky. We often hear frightening tales from our members of the older or bigger ferret beating up the new or younger one. Some of this interaction is normal—this is the way ferrets relate. However, they must be closely watched to be sure no one is hurt. We suggest a gradual introduction—let each ferret spend one night on a towel in cages placed in separate rooms. The next day switch towels to let them get used to each other's scent. The following day take both ferrets to "neutral" territory—away from either's cage, food or toys. Speaking softly, pet them both and give them treats—this will make them associate pleasant things with being together. Let them stay together until the fighting

The bottom of this cage has been lined with wood shavings to absorb the ferrets' waste material. Wood shavings are suitable for this purpose, but they should not be used for bedding.

Play biting is normal for young ferrets.

becomes too rough—then separate them. Continue the sessions with love and treats until you are 100% sure they will not hurt each other when left alone. This process can take as little as three days or as much as several weeks. However, be assured in time they will be inseparable.

Several months after we adopted Furry, we heard of a ferret who had been given as a gift to a woman who had no desire for a pet. She ignored the ferret and sometimes neglected to feed her. We decided we could not let this happen while we were able to house another ferret, so we asked for her. The night our new little girl, Heather, arrived was utter chaos. Although she weighed only one pound, she was a tigress. After she bit through Jay's fingernail we put her in her new cage to calm down. Although she could not get near Furry, he was terrified and started to claw at the wall to try to escape. My feeling was that this would never work; my husband, who was more patient, felt things would be fine. We found that she was in heat and took steps to correct that situation. After four months of good treatment, regular food, and lots of affection, she became the most loving, cuddling pet anyone could ever wish for. All she wanted in life was to be held or carried around (an easy thing to do since she never weighed more than 1-$^{1}/_{4}$ pounds). She and Furry became inseparable—when they slept it was almost impossible to determine where one began and the other left off. It's hard to say who benefited more by this adoption—Heather, Furry or us. I think it was a win-win-win situation all around.

One of our members reported, "It took one whole year for my ferrets to become friendly. A little over two years ago I adopted a $2^{1}/_{2}$ year old, spayed and descented female ferret from the Humane Society. When I

These ferrets are being introduced for the first time. Use common sense when bringing a new ferret into a household that already has a resident ferret; don't expect them to instantly become friends.

Ferrets are small, quiet pets, which makes them particularly suited for living in an apartment.

tried to introduce her to my year-old female (also spayed and descented), the new ferret attacked her immediately and tried to seriously harm her. After several weeks of constant attacking by the older ferret, I was advised by my veterinarian to let them fight it out. I couldn't do it because I knew that the older ferret could kill my little one.

For one year I had to keep them separated at all times even though they would bite and claw the cages because they wanted to get out. I felt guilty that I had brought an intruder into the house, and my younger ferret's life was totally disrupted. I refused to take the older ferret back to the Humane Society because I had made a commitment to take care of her. One day a miracle occurred—I decided to let them out at the same time. As usual, the older ferret attacked, but this time the younger one refused to take any more abuse and attacked in return. The older ferret backed off immediately. It took several more months before I lost the fear that the older one would hurt the younger one and I watched constantly for any signs of trouble."

Q. One of my ferret's "vices" is to scratch in the dirt around my houseplants, which makes a mess. Any ideas?

A. Since ferret droppings do not have an odor, remove some feces from the litter pan and place on top of the dirt around the plants.

Q. I am in the process of seeking a new apartment and am having trouble finding a suitable one that will accept a pet.

A. A no-pet clause in leases saves landlords the trouble of dealing with tenants on an individual basis. If nobody is allowed to have a pet, that's one less hassle for the landlord. Certainly a landlord might be justified in not wanting a 125-pound dog in a three-room apartment, but by declaring "no pets allowed," such pets as ferrets are also excluded. When shopping for a new home, we suggest you take along a picture of your ferret. It's hard to refuse permission to have such a small, quiet pet. If the landlord verbally agrees you can have your ferret, get it in writing in your lease.

One of our members lived with his ferret in an exclusive

apartment house that had a strict "no pet" rule. When asked periodically if he had any pets, he would reply, "Only Goldfish". The ruse worked and was only discovered on the day our member and his ferret "Goldfish" moved to an apartment friendlier to little pets.

Q. If I am unable to take my ferret on vacation with me, what should I do? A. If you are taking your ferret to a boarding facility (kennel), check the place out. Is it clean and odor free? Is the staff friendly and attentive to animals? Will you be permitted to provide your pet's favorite food to avoid feeding problems? If the facility is not affiliated with a veterinarian, be sure to leave the name and phone number of your pet's vet and provide a copy of his/her medical records.

If your pet will remain at home, be sure to have someone come in at least twice a day to check on him/her, empty the litter pan, provide food and fresh water and companionship. The ferret should be confined to the cage unless the "sitter" is present. Your sitter should know the signs of illness in a ferret. Leave a book on ferret care for reference along with the name and phone number of your veterinarian.

Also it is important to consider preparing for an unplanned absence from home. Bad weather, automotive failure, ill health or accident may prevent us from returning home as planned. It is a good idea to give a house key to a trusted neighbor, instruct that person on ferret care and leave a note in your wallet that the neighbor should be notified in case of emergency. Since ferrets are so small, it takes very little time for them to dehydrate due to lack of water and become extremely hungry. This plan takes only a few minutes but will do much to relieve your mind if you should be unexpectedly detained.

A ten-month-old sable female exploring the back yard.

Q. Lately the ferret I've had for three years has started to "act out". Is she misbehaving deliberately just to annoy us?

A. This is probably not the case. Your pet may not know what he/she is doing is wrong. It may be necessary to patiently "inform" your ferret that what is being done constitutes misbehavior. This can be done by saying "no" in a firm voice, removing him/her from the place of misbehavior or squirting in the face with water from a plant mister. Never give your pet a treat or playtime sooner than 10 minutes after the misbehavior.

There are other reasons for misbehavior: fear, loneliness, boredom and illness among them. When frightened, some ferrets will hide under their blankets; others will misbehave. If your pet is misbehaving while ill, this is understandable, and, if possible, permit the ferret to do as he/she wishes. Usually as soon as the pet is feeling better, misbehavior will stop.

If your ferret consistently knocks over an item in its reach, it's usually simple to move the item. Be sure your pet has space and time enough to exercise. Play uses a lot of ferret energy and leaves less for misdeeds.

Q. In the spring my ferret seems to shed. Is this normal?

Don't worry—your pet is just preparing for warm weather. To speed the process, you might use a medium-firm hair brush or remove the shed hair with a bath. Your pet might scratch a bit more than usual during shedding time. The new coat will be slightly shorter and lighter.

Silver-mitt youngster. Silver-mitts are among the most popular color varieties in ferrets.

Q. What should I do if my ferret gets lost?

A. First determine if there is any way he/she could have gotten out of the house. If there is no way that could happen, use the squeaker toy (repeatedly if necessary). If the ferret doesn't respond, try listening for scratching sounds in the walls or behind furniture, in case it is trapped somewhere. Keep food and water in the usual place—your ferret may be napping and will come out for nourishment when ready.

If you believe the ferret is outdoors, again use the squeaker repeatedly. Alert all neighbors and place signs in the neighborhood (including a picture of your pet). Often a good source of information is the mail carrier. He/she can spread the word and might hear of a ferret found in the area. Place dishes of your ferret's food and water by doors and place old towels or your ferret's favorite blanket by the food and water dishes. Check with local Humane Societies and tell area veterinarians. Call the FFC at (412) 322-1161—we will note your phone number and (as has often happened) may be able to give you the number of someone who reported finding a ferret.

To prevent problems whenever you are having guests in the house, it is a good idea to introduce your ferret to them and warn them of any dangers they could unknowingly cause for your pets. Warn them not to leave doors or windows open. When helping with household chores, make sure that your pet doesn't get thrown out with the trash!

Ferrets are active little pets, but they do need ample sleep to re-energize themselves.

SUGGESTED READING

T. F. H. Publications offers a comprehensive selection of books dealing with ferrets. A selection of titles is presented below; they and many other works are available from your local pet shop.

PS-754
64 pages
22 color photos

KW-074
128 pages
Over 90 color photos

YF-105
32 pages
16 color photos

TU-014
64 pages
Over 60 color photos

SK-009
64 pages
Over 50 color photos

TS-106
160 pages
Over 90 color photos